Dyslexia and Traumatic Experiences

Beiträge zur Pädagogischen und
Rehabilitationspsychologie

Studies in Educational and
Rehabilitation Psychology

Herausgegeben von / Edited by Evelin Witruk

Bd./Vol. 7

Evelin Witruk / Shally Novita / Yumi Lee /
Dian Sari Utami (eds.)

Dyslexia and Traumatic Experiences

Bibliographic Information published by the Deutsche Nationalbibliothek
The Deutsche Nationalbibliothek lists this publication in the Deutsche
Nationalbibliografie; detailed bibliographic data is available in the internet at
http://dnb.d-nb.de.

Library of Congress Cataloging-in-Publication Data
Names: Witruk, Evelin, editor. | Novita, Shally, 1983- editor. | Lee, Yumi,
1980- editor. | Utami, Dian Sari, editor.
Title: Dyslexia and traumatic experiences / Evelin Witruk, Shally Novita, Yumi
Lee, Dian Sari Utami, (eds).
Description: Frankfurt am Main ; New York : Peter Lang, 2016. | Series: Studies
in educational and rehabilitation psychology ; Vol. 7
Identifiers: LCCN 2016003873 | ISBN 9783631661154
Subjects: LCSH: Dyslexia--Psychological aspects. | Post-traumatic stress
disorder in children--Treatment.
Classification: LCC RC394.W6 D954 2016 | DDC 616.85/53--dc23 LC record
available at http://lccn.loc.gov/2016003873

ISSN 1865-083X
ISBN 978-3-631-66115-4 (Print)
E-ISBN 978-3-653-05604-4 (E-Book)
DOI 10.3726/978-3-653-05604-4

© Peter Lang GmbH
Internationaler Verlag der Wissenschaften
Frankfurt am Main 2016
All rights reserved.
PL Academic Research is an Imprint of Peter Lang GmbH.

Peter Lang – Frankfurt am Main · Bern · Bruxelles · New York ·
Oxford · Warszawa · Wien

This publication has been peer reviewed.

www.peterlang.com

Table of Contents

Evelin Witruk, Shally Novita, Yumi Lee, & Dian Sari Utami

(University of Leipzig, Institute of Psychology)

Preface

This book is the seventh volume in the series "Studies in Educational and Rehabilitation Psychology". It contains selected contributions from the international conference "Dyslexia and Traumatic Experiences" organized by the team members of Educational and Rehabilitation Psychology, Institute of Psychology at the University of Leipzig. It took place on 5 and 6th of December 2014 in the University of Leipzig, Germany.

The purpose of this book is to strive towards fostering a scientific exchange that promotes emergence of synergy effects and scientific progress. The authors of the book articles are from Indonesia, Sri Lanka, Morocco, Sudan, South Africa, South Korea, Iran, China, Portugal, and Germany. The interdisciplinary character of this book is representing in contributions of scientists from psychology, special education, linguistics, genetics, and neuropsychology.

The main topics of the book are structured in four chapters. They are related to dyslexia with some new perspectives on this old phenomenon, traumatic experiences, intervention methods, and some special methodical problems, particularly in qualitative research methods.

The authors of the book articles, the participants of the workshop, as well as the editors were very grateful for the sponsorship of the DAAD for scientists from Sri Lanka.

Evelin Witruk   Leipzig, March 2016
Shally Novita
Yumi Lee
Dian Sari Utami

Chapter 1
Dyslexia

Evelin Witruk

University of Leipzig, Germany

Dyslexia – New Perspectives on an Old Phenomenon

Abstract. The article aims to the question what is new until the last two decades in the dyslexia research and in the assessment. Some new aspects and some lines of progress will be discussed regarding the genetic basis of dyslexia, the hemispheric dominance, and the visual-spatial abilities of dyslexic individuals in different ages.

Keywords: genetic basis of Dyslexia, hemispheric laterality, visual-spatial abilities.

1 From family and twin studies to the analysis of the genetic code

Family and twin studies indicated over decades a strong hereditary disposition of dyslexia. The studies showed that 40 % of the siblings and parents of a dyslexic were also having dyslexia (Grimm, 2001; Wilcke & Boltze, 2010). In twin studies, the genetic determination of dyslexia was highly estimated (about 60 %; Olson, Forsberg, & Wise, 1994). But the critical point regarding these studies is that the non-measured impact of family members, of cultural and natural environment were not be considered.

Linkage studies are a way to narrow the genomic region, where relevant disease genes are expected. Several genes have been linked to dyslexia, including DCDC2 and KIAA0319 on chromosome 6 and DYX1C1 on chromosome 15 (e. g., Grigorenko, et al., 1997). But, these findings are not always replicated. Molecular studies have linked several forms of dyslexia and different cognitive processes to genetic markers. However, no single gene is definitively implicated in dyslexia. Linkage analysis showed until now that at least nine different chromosomal regions could be identified where several disease genes are suspected. Those regions are connected with dyslexia, and are called DYX regions (Witruk & Wilcke, 2010).

Association studies focused on genes previously identified in linkage studies as potential candidates and compared different populations (i.e., dyslexics vs. controls). Most relevant are the analysis of SNP (Single Nucleotide Polymorphism), which means that a single base at a certain position in the genome is different in some individuals, and that these individuals comprise at least 1 % of the population.

Wysocka, Lipowska, and Kilikowska (2010) could show that dyslexia "seems to be a complex trait determined by number of genes, with small to moderate effects on the specific phenotype, involving various factors such as heterogeneity, incomplete penetrance, phenocopy, or oligogenic inheritance. Based on combined linkage and association analysis using both qualitative and quantitative phenotypes, the multiple regions (DYX1-DYX9) on chromosomes 1, 2, 3, 6, 11, 13, 15 and 18 have been reported likely to contain genes contributing to dyslexia. Most recently, four candidate genes (DYX1C1, KIAA0319, DCDC2, ROBO1) have been identified as associated with dyslexia" (Wysocka et al., 2010). Therefore, it is possible that one person has some genetic risk variants and some protective variants that compensate each other. Depending on the number and type of genetic risk variants, a mild, moderate or severe type of dyslexia is developed (Witruk & Wilcke, 2010).

2 From the assumption of left handedness to hemispheric laterality profiles

The former assumption of left handedness as a characteristic of dyslexic individuals could not be confirmed in the last decades. Several empirical studies found a weak, combined laterality, and hemispheric coordination problems among dyslexic children. Larsen, Höien, Lundberg, and Ödegaard (1990) found a reduction of the usual asymmetry of the left and right Planum Temporale as well as a high correlation of mixed handedness and phonological disorders. Stein (1994) explained dyslexia by the impaired magnocellular functions and the impaired hemispheric specialization and lateralization. Sebastian and Yasin (2008) showed in a Mismatch Negativity experiment with compensated dyslexic adults that the lateralization of the auditory system can be less specialized as a result of impaired hemisphere dominance.

Our research investigated the laterality profiles in dyslexic and normal-reading children in connection with their phonological awareness (Schulz, 2013), their intelligence profiles and reading and spelling performances (Unger, 2007). Two studies of laterality effects (hemispheric dominance effects) on hands, eyes, and legs in dyslexic children were discussed. It could be confirmed our assumption of weak and combined hemispheric laterality in dyslexic children and its motoric and sensory behavioral consequences on preferences of hands, eyes, and legs in dyslexic children. The individual laterality profiles were compared between dyslexic and normal reading children on the basis of discriminate and cluster analysis. The results show a dependency on dyslexia, gender, and a correlation to the phonological awareness.

3 Visual-spatial abilities: Deficits versus strengths?

The beginning of dyslexia research is connected with the assumption of special, and strong visual impairments in the sense of "congenital word blindness" (Orton, 1925) and the "Raum-Lage-Labilität" (Schenk-Danzinger, 1991). In several studies, visual deficits were found in dyslexic individuals. In most of the studies including dyslexic children visual deficits could be confirmed, such as Lipowska, Czaplewska, and Wysocka (2011), whereas other studies (e.g., Graeve, 1997), found no significant differences or could show advantages in the dyslexic individuals (Witruk, 2011, 2015). Therefore the question can be generated regarding the compensation effects during the life span of dyslexic individuals or visual-spatial strengths which can be connected with dyslexia.

Deficits in script acquisition can be the expression of a global, holistic processing style which can have advantages within several other visual requirements compared to reading and writing (Brunswick, Martin, & Marzano, 2010; Károlyi, Winner, Gray, & Sherman, 2003). This global, holistic processing style can be based on the reduced hemispheric asymmetry (Larsen, Höien, Lundberg, & Ödegaard, 1990). Our research is caused by the controversial findings regarding visual-spatial abilities in dyslexic individuals and the clear link to gender dependency of these abilities.

In three experiments, we used visual tasks which can be solved by different cognitive processing styles. In contrast to the analytic processing style, the global, holistic processing style is possible with assumed advantages for the accuracy and the reaction speed. We asked like Tafti, Hameedy, and Baghal (2009) and Wolff and Lundberg (2002) about the advantages in the sense of talents or compensation products of dyslexic individuals regarding visual-spatial abilities. We assumed that compensation products are developing over the school time and are completed in the adolescence. Therefore, we integrated dyslexic and control individuals from different age groups (children with a mean age of 10.26 years, adolescents with a mean age of 17.16 years and young adults with a mean age of 23.04 years) and from different orthographic background (Cantonese ideophonetic, Arabic segmental, and German alphabetic script). The results could confirm our assumption of visual-spatial advantages in dyslexic individuals in dependency of gender, age group, and the type of orthography. The advantages were clear in the group of adolescents and therefore they can be interpreted as compensation products (Witruk, 2015).

One of the conclusions of these findings led to the development and application of virtual realities for the assessment and treatment of dyslexic individuals on the basis of their visual-spatial strengths. Attree, Turner, and Cowell (2009) could show that the visual-spatial strengths of dyslexics are to observe in the age of adolescents, not only on the basis of traditional paper and pencil test (here used British Ability Scale, BAS II), but also on the basis of virtual reality tasks. They constructed a virtual reality test by using Superscape VRT software and could show significant better spatial recognition memory performances among dyslexic adolescents comparing with a control group. The authors conclude that the learning process of dyslexic children should integrate their strengths from the beginning. Using techniques that help them to learn through their strengths can enable successful learning. On this way they expect prevention against strong primary (failures in reading and/or writing) and secondary symptoms (e.g., anxiety, low self-esteem, and low motivation) of dyslexic individuals.

4 Affiliation

Prof. Dr. Evelin Witruk
Institution: University of Leipzig, Educational and Rehabilitation Psychology
Address: Neumarkt 9–19, 04109 Leipzig, Germany
E-mail: witruk@uni-leipzig.de

5 References

Attree, E. A., Turner, M. J., & Cowell, N. (2009). A Virtual Reality Test Identifies the Visuospatial Strengths of Adolescents with Dyslexia. *Cyber Psychology & Behavior, 12*(2), 163–168.

Brunswick, N., Martin, G. N., & Marzano, L. (2010). Visuospatial superiority in developmental dyslexia: Myth or reality? *Learning and Individual Differences, 20*(5), 421–426.

Graeve, J. (1997). *Legastheniespezifik visueller sukzessiver Vergleichsleistungen bei verschiedenen Materialarten*. (Unpublished master thesis). University of Leipzig, Leipzig, Germany.

Grigorenko, E. L., Wood, F. B., Meyer, M. S., Hart, L. A., Speed, W. C., Shuster, A., & Pauls, D. L. (1997). Susceptibility loci for distinct components of developmental dyslexia on chromosome 6 and 15. *American Journal of Human Genetics, 60*, 27–39.

Grimm, T. (2001). Genetische Ursachen der Legasthenie. In Deutsche Gesellschaft für das hochbegabte Kind e. V. (Hrsg.), *Hochbegabte Kinder in Schule und Gesellschaft*. (S. 37–43). Münster: LIT Verlag.

Károlyi, C. von, Winner, E., Gray, W., & Sherman, G. F. (2003). Dyslexia linked to talent: Global visual-spatial ability. *Brain and Language, 85*, 427–431.

Larsen, J. P., Höien, T., Lundberg, J., & Ödegaard, H. (1990). MRI Evaluation of the Size and Symmetry of the Planum Temporale in Adolescents with Developmental Dyslexia. *Brain and Language, 39*, 289–301.

Lipowska, M., Czaplewska, E., & Wysocka, A. (2011). Visuospatial deficits of dyslexic children. *Medical Science Monitor, 17*(4), 216–221.

Olson, R. K., Forsberg, H., & Wise, B. (1994). Genes, environment, and development of orthographic skills. In V. W. Berninger (Ed.), *The varieties of orthographic knowledge I: Theoretical and developmental issues* (pp. 27–71). Dordrecht: Kluwer.

Orton, S. T. (1925). 'Word-blindness' in school children. *Archives of Neurology and Psychiatry, 14*, 285–516.

Schenk-Danzinger, L. (1991). *Legasthenie*. München: Ernst Reinhardt.

Schulz, S. (2013). *Legasthenie und Lateralitat unter Einbeziehung der phonologischen Bewusstheit.* (Unpublished bachelor thesis). University of Leipzig, Leipzig, Germany.

Sebastian, C., & Yasin, I. (2008). Speech versus tone processing in compensated dyslexia: discrimination and lateralization with a dichotic mismatch negativity (MMN) paradigm. *International Journal of Psychophysiology, 70*(2), 115–126.

Stein, J. F. (1994). Developmental dyslexia, neural timing and hemispheric lateralisation. *International Journal of Psychophysiology, 18*(3), 241–249.

Tafti, M. A., Hameedy, M. A., & Baghal, N. M. (2009). Dyslexia, a deficit or a difference: Comparing the creativity and memory skills of dyslexic and nondyslexic students in Iran. *Social Behavior and Personality: an international journal, 37*(8), 1009–1016.

Unger, P. (2007). *Legasthenie und Lateralitat.* (Unpublished master thesis). University of Leipzig, Leipzig, Germany. .

Wilcke, A., & Boltze, J. (2010). Genetische Grundlagen der Legasthenie. In E. Witruk, D. Riha, A. Teichert, N. Haase & M. Stueck (Hrsg.), *Learning, Adjustment and Stress Disorders. Band. 1: Beiträge zur Pädagogischen und Rehabilitationspsychologie* (S. 59–82). Frankfurt/Main: Peter Lang.

Witruk, E. (2015, December). Dyslexia, compensation effects, and coping. *Paper presented at the International Workshop: Dyslexia and Coping Behaviour.* Leipzig, Germany.

Witruk, E., & Wilcke, A. (2010). Dyslexia – an overview of assessment and treatment methods – with special reference of genetic basics. *Buletin Psikologi, 18*, 69–90.

Witruk, E. (2011). Assessment and treatment of dyslexia – an overview. *Ad verba liberorum: Linguistics & Pedagogy & Psychology, 3*, 4–18.

Wolff, U., & Lundberg, I. (2002). The prevalence of dyslexia among art students. *Dyslexia, 8,* 34–42.

Wysocka, A., Lipowska, M., & Kilikowska, A. (2010). Genetics in solving dyslexia puzzles: The Overview. *Acta Neuropsychologica, 8,* 315–331.

Shally Novita[1] & Evelin Witruk[2]

[1] Leibniz Institute for Educational Trajectories (LifBi), Germany

[2] University of Leipzig, Germany

Emotional Consequences of Children with Dyslexia: An Overview from a Cross-cultural Perspective

Abstract. For more than a decade, the emotional impacts of dyslexia on the lives of individuals have been studied from different aims and perspectives. However, most studies on this topic have been conducted in a single cultural context. This study investigated the cross-cultural differences between children with and without dyslexia, specifically in respect of their anxiety and self-esteem profiles. A total of 124 children with and without dyslexia from Germany and Indonesia participated in this study. They were comparable in age (8–11 year olds), school grade (third and fourth grade) and IQ (> 73). All children were administered an IQ test (CFT-20R) and completed two questionnaires (i.e., Spence's Children Anxiety Scale and The General List of Self-esteem for Children and Adolescent). This study cannot provide significant results for hypotheses proposed. However, weak-medium effect sizes were reported for the effect of dyslexia on anxiety ($d = -.21$), dyslexia on self-esteem ($d = .34$) and different anxiety levels of German and Indonesian children ($d = -.31$).

Keywords: anxiety, self-esteem, dyslexia, cross-culture.

1 Introduction

Dyslexia is a specific learning disability that has a neurobiological origin (Lyon, Shaywitz, & Shaywitz, 2003) and has no significant relationship with IQ (Witruk & Eichhorn, 2012). According to Betz and Breuninger (1993), children with dyslexia may experience what they refer to as the four stages of a *virtuous circle of learning disorder*.

2 Theory

2.1 Dyslexia and emotional consequences

The studies on the role of emotion in academic and reading-writing performance reported that individuals with dyslexia have disadvantages in respect of their anxiety level (Caroll & Iles, 2006; Nelson & Harwood, 2011). In general, children with learning difficulty reported a lower score of positive well-being, were unhappier and more anxious than their peers without similar difficulties (Casey, Levy, Brown, & Brooks-Gunn, 1992). A meta-analysis by Nelson & Harwood (2011) also reported a statistically significant medium of effect size ($d = .61$) on anxiety symptom of school age children with learning disabilities. More specifically, studies reported that children and teenagers with dyslexia have a lower level of perceived scholastic competence (Frederickson & Jacobs, 2001), lower level of achievement, effort investment, academic efficacy, sense of coherence, positive mood, and hope (Lackaye & Margalit, 2006), and have more academic, social and psychological problems (Vigilante & Dane, 1991) than their peers without similar difficulty. Over a prolonged period, children who showed high levels of anxiety could have negative educational outcomes such as failure to complete high school or college (Ameringen, Mancini, & Farvolden, 2003; Kessler, Foster, Saunders, & Stang, 1995).

2.2 Culture and anxiety

Hofstede (2001) introduced one dimension called the uncertainty avoidance index as one important factor for investigating anxiety in a cross-cultural context. This dimension shows how culture is dealing with an ambiguous situation, and it is strongly related to anxiety (Hofstede, Hofstede, & Minkov, 2010). This result gives strong evidence that culture is an important factor that should be considered in anxiety research.

2.3 Culture and self-esteem

According to Tsai, Ling, and Lee (2001) in the individualistic culture, people tend to see the self as separate from others. They argued that in this culture, individuals are encouraged to express their uniqueness by engaging in self-enhancement strategy (i.e., presentation of the self as superior to others). On the other hand, collectivistic culture tends to see the self as part of others and, therefore, encourages their member to maintain the harmony of an interpersonal relationship through self-effacement strategy (i.e., presentation of the self as inferior to others). As a result, it is widely assumed, that Westerners view themselves more positively than Asians (Brown & Cai, 2010).

2.4 Hypotheses

Hypothesis 1: children with dyslexia are more vulnerable to emotional consequences such as low self-esteem and high anxiety compared to children without dyslexia.

Hypothesis 2: due to cultural differences, Indonesian and German children develop different anxiety and self-esteem profiles.

3 Method

3.1 Sample

A total of 124 children from Indonesia and Germany participated in this study. The ratio between children with and without dyslexia was 64 (M_{age} = 8.86) to 60 (M_{age} = 9.23). They were comparable in IQ ($M_{IQnondys}$ = 101.09, M_{IQdys} = 97.72. t = 1.45, p = 1.15), gender (boys = 62, girls = 62), and were assigned to either third (n = 57) or fourth year at school (n = 67).

The Indonesian group was represented by the following characteristics: 29 children with dyslexia (M_{age} = 8.93) and 35 children without dyslexia (M_{age} = 8.49), 34 third and 30 fourth year pupils, 28 boys and 36 girls. The German group consisted of 31 children with dyslexia (M_{age} = 9.52) and 29 children without dyslexia (M_{age} = 9.31), in year three (n = 23) and year four (n = 37), represented by both genders (boys, n = 34 and girls, n = 26). The children with dyslexia were diagnosed by qualified psychologists in both countries, and all of the children without dyslexia had no history of learning difficulties.

3.2 Measurement tools

Measurement tools that were used in this study are: General List of Self-esteem for Children and Adolescent (Schauder, 1991), Spence Children's Anxiety Scale (Spence, 1998), Culture Fair Intelligence Test 20 Revision ([CFT-20R] see also: Weiss, 2006).

4 Results

Multiple regressions were conducted to test the hypothesis. Table 1 reports slopes, R-values and effect size (Cohen d) of the conducted analysis.

Table 1. Results of multiple regressions

	Slope	R	d
Dyslexia and anxiety	.13	.20	-.24
Country and anxiety	.16		-.31
Dyslexia and self-esteem	-.01	.17	.34
Country and self-esteem	-.17		-.01

Note. Group coded: 1 = children without dyslexia, 2 = children with dyslexia. Country coded: 1 = German, 2 = Indonesia. Interpretation of *d* values: .2 = weak, .5 = medium, .8 = large effect (Cohen, 1988). Negative effect sizes reflect that second group has higher mean than first group.

No significant effects were found in either analysis. Children with dyslexia have relatively similar anxiety and self-esteem profiles compared to children without dyslexia. Country is also not regarded as a significant predictor for anxiety and self-esteem profiles of children in age groups between 8–11 years.

5 Discussion

This current study can neither support the assumption of emotional vulnerabilities of children with dyslexia nor the different anxiety and self-esteem profiles of children from different countries. However, weak-medium effect sizes were found for the effect of country and dyslexia on anxiety as well as the effect of dyslexia on self-esteem. The contradiction of significant test and effect size analysis is recognized as a result of power issue, which should be addressed in the further cross-cultural study.

6 Affiliations

Dr. Shally Novita
Institution: Leibniz Institute for Educational Trajectories (LifBi)
Address: Wilhelmsplatz 3, 96047, Bamberg, Germany
E-mail: shally.novita@lifbi.de

Prof. Dr. Evelin Witruk
Institution: University of Leipzig, Educational and Rehabilitation Psychology
Address: Neuemarkt 9–19, 04103 Leipzig, Germany
E-mail: witruk@uni-leipzig.de

7 References

Ameringen, M. V., Mancini, C., & Farvolden, P. (2003). The impact of anxiety disorders on educational achievement. *Journal of Anxiety Disorders, 17*, 561–571.

Betz, D., & Breuninger, H. (1993). *Teufelkreis Lernstoerungen. Theoretische Grundlegung und Standardprogramm* (3rd ed.). Munich: Beltz Psychologie Verlags Union.

Brown, J. D., & Cai, H. (2010). Self-esteem and trait importance moderate cultural differences in self-evaluations. *Journal of Cross-Cultural Psychology, 41*, 116–123.

Caroll, J. M., & Iles, J. E. (2006). An assessment of anxiety level of dyslexic students in higher education. *British Journal of Educational Psychology, 76*, 651–662.

Casey, R., Levy, S. E., Brown, K., & Brooks-Gunn, J. (1992). Impaired emotional healt in children with mild reading disability. *Developmental and Behavioural Pediatrics, 13*, 256–260.

Cohen, J. (1988). *Statistical Power Analysis for The Behavioral Sciences.* New Jersey: Lawrence Erlbaum Associates.

Frederickson, N., & Jacobs, S. (2001). Controllability attributions for academic performance and the perceived scholastic competence, global self-worth and achievement of children with dyslexia. *School Psychology International, 22*, 401–416.

Hofstede, G. (2001). *Culture's Consequences: Comparing Values, Behaviors, Institutions, and Organizations Across Nations* (2nd ed.). Thousand Oaks: Sage Publication.

Hofstede, G., Hofstede, G. J., & Minkov, M. (2010). *Cultures and Organizations. Software of The Mind* (3rd ed.). New York: McGraw Hill.

Kessler, R. C., Foster, C. L., Saunders, W. B., & Stang, P. E. (1995). Social consequences of psychiatric disorders, I: Educational attainment. *Am J Psychiatry, 152*, 1036–1032.

Lackaye, T. D., & Margalit, M. (2006). Comparison of achievement, effort, and self-perception among students with learning disabilities and their peers from different achivment groups. *Journal of Learning Disabilities, 39*, 432–446.

Lyon, G. R., Shaywitz, S. E., & Shaywitz, B. A. (2003). Defining Dyslexia, Comorbidity, Teachers' Knowledge of Language and Writings. *Annals of Dyslexia,* (53), 1–15.

Nelson, J. M., & Harwood, H. (2011). Learning disabilities and anxiety: A meta analysis. *Journal of Learning Disabilities, 44*, 3–17.

Schauder, T. (1991). *Die Aussagen-Liste zum Selbstwertgefühl für Kinder und Jugendliche.* Weinheim: Beltz Test.

Spence, S. H. (1998). A measure of anxiety symptoms among children. *Behaviour-Research and Therapy, 36*, 545–566.

Tsai, J. L., Ling, Y. W., & Lee, P. A. (2001). Cultural predictors of self-esteem: a study of chinese american female and male young adults. *Cultural Diversity and Ethnic Minority Psychology, 7*(3), 284–297.

Vigilante, F. W., & Dane, E. (1991). Teenage dyslexia: stuerm und drang. *Child and Adolescent Social Work, 8*(6), 515–523.

Weiss, R. H. (2006). *CFT-20R. Grundintelligenztest Skala 2.* Goettingen: Hogrefe.

Witruk, E., & Eichhorn, R. (2012). An overview of assessment and treatment methods of dyslexia with special references to emotional and behavioural problems. *Scientific Journal of Saratov University, 10*, 142–154.

Regine Eichhorn

University of Leipzig, Germany

Secondary Symptoms and Compensation – Mechanisms of Dyslexic Children

Abstract. This longitudinal-study will continue the investigation of the development of secondary symptoms in dyslexic children, the impact of dyslexia on the self-esteem, anxiety parameters, motivation aspects and behavioral components. The focus of this study is the investigation of the benefit of special dyslexic rehabilitative classes in comparison to integrative classes in Germany. The study includes four measurement points and is still in process. The third measurement point is finished. The research questions are: How do the scholastic surrounding conditions influence the well- being of the dyslexic children and the perceptions and evaluations of the teachers and parents concerning the behavior and emotional expressions of the children? Which reaching method has the best preventative effect? The current results give first evidence for the positive effect of special rehabilitative classes for dyslexic children on their subjective well-being.

Keywords: special rehabilitative dyslexic classes, integrative classes, longitudinal study.

1 Introduction

Reading and writing are the most important competences to take part in the society. Having a handicap in these sectors/fields can cause massive consequences. Especially children with dyslexia are confronted with huge school and social problems. The association of children with learning deficits or disorders is very different in each state of Germany. Special rehabilitative dyslexic classes were established in the easterly states of Germany. But for the last 2–3 years these classes were removed in the eastern part of Germany and still implemented in the state of Saxony. In the states without special dyslexic classes the schooling advancement implied some remedial lessons and "disadvantage adjustment" (e.g., more time to work on reading and writing tests or tasks). And depending on the financial possibilities of the parents there are a lot of private learning institutes where the children can get special help and support. In the state of Saxony, where the special dyslexic classes still exist, children who show massive problems to learn reading and writing during the second grade, are doing a special diagnose period. When the child is diagnosed with dyslexia, he will go for two years in special dyslexic classes. These classes are small groups of 8–12 children taught by special educated teachers. After the two years the child returns to regular classes for one year and go back to normal school for the fourth grade. In the states without special dyslexic classes, all children have lessons together, independent from their handicap. The main goal in those states is inclusion. The difference between the two teaching methods has to be analyzed considering the background of the theory of the development of secondary symptoms and the past results of Eichhorn (2010, 2012), which showed the positive effects of special dyslexic classes for the well-being of the children.

2 Theory

The theoretical background of the contemporary study is the theory of Betz and Breuninger "Teufelskreis Lernstoerungen" (1998). The authors describe the development of the secondary symptoms of the dyslexic children. This term is based on the model of Valtin (1989). It distinguishes between etiological and phenomenological level. There are primary and secondary causes for dyslexia, like genes and deficits in basic competences (e.g., working memory) and which entail primary symptoms (reading and writing problems) and this involving again secondary symptoms (behavioral and emotional problems). In their "vicious cycle" Betz and Breuninger (1998) present four steps for the consequences of a learning deficit. Destabilize self-esteem, leads to reduce learning motivation and increase anxiety of the affected child, because of the self- attribution of the learning deficit (without knowing that it is a handicap). Further interactions with the social environment and misunderstandings from the parents and the teachers reinforce the self-esteem problems. So the child develops behavioral and emotional conspicuities (e.g., school avoidance, problems in other subjects). If it has come to the last stadium in the vicious cycle, all described processes strengthen and the child as well as his teachers and parents do not expect any scholastic success anymore. The theory of Betz and Breuninger (1998) does not consider that the special dyslexic classes can have an influence or any other preventative effect on the children who have secondary symptoms.

3 Method

3.1 Sample

The first data collection of the study includes the statistics from 207 participants (112 female, 95 male). 44 children diagnosed with dyslexia were taught in special dyslexic classes. 8 children also with dyslexia were taught in integrative classes and 155 children without dyslexia (control group) took part. 33 children were diagnosed with a psychological or physiological disorder and 9 of them had an ADHD.

The first data collection includes results from 192 children questionnaires, 174 teachers' questionnaires and 189 parental questionnaires. The second data collection includes 160 children questionnaires (72 female, 88 male), 146 teachers' and 134 parental ones. The third data collection includes statistics from 143 children (63 female, 81 male). 43 children were taught in special dyslexic classes, 8 children in integrative classes and 72 were in the control group.

The results include 143 teachers and 130 parental questionnaires. There are 123 children who took part from the first until the third data collection. There is a high dropout but also new participants in the second and third data collection. Overall, three schools with special rehabilitative dyslexic classes and eight other schools, with regular classes, participated to the study.

3.2 Measurement tools

A longitudinal study, which includes four data collections, was chosen to investigate the development of dyslexic children secondary symptoms. The first data collection was in autumn 2013 with the beginning of the school year. The dyslexic children began their treatment in special rehabilitative classes. The control group and the integrative one started it at the beginning of the second grade. So the study includes two experimental groups (first: children with dyslexia in special dyslexic classes, second: children with dyslexia in integrative classes), and one control group (children without dyslexia). When the data was collected for the second time, it was during the second half of the second school year (for the control and integrative class group) and the second half of the first special dyslexic class year. When the data was collected for the third time it was during the third grade respectively in the second dyslexic class year.

The fourth and the last data collection will be conducted during the beginning of the fourth grade, when the dyslexic class children return to regular classes and treatment.

In each data collection 5 questionnaires were used. Three were for the self-evaluation of the children, to measure self-esteem, anxiety, learning and achievement motivation. One was used for the parents and one for the teachers, to measure their evaluations concerning the behavior, the competences and the internalizing and externalizing symptoms which they observed in their children or students.

The children got child-friendly questionnaires which got another layout to make the reading easier for the children and to increase the answer motivation.

The teachers were instructed how to do the questionnaires with the children. In the first data collection all questionnaires were taken in their original length. For the following data collections the questionnaires were shortened to reduce the time costs for the teachers and to keep up the participation and prevent drop-outs. Because of the disposed dyslexic diagnose and further disposed special dyslexic classes in Saxony- Anhalt and to find out which child has dyslexia, there was an individual testing of suspicious children with common tests.

Questionnaire for the children. To measure the self-evaluated well-being, the "Angstfragebogen fuer Schueler (AFS)" (Wieczerkowski, Nickel, Janowski, Fittkau, & Rauer, 1973) was used. It includes four scales: one to measure anxiety in general, another also to measure anxiety but during exams, one scale to measure school aversion and a fourth one to measure social desirability. Then, "Skala zur Erfassung der Lern-Leistungsmotivation (SELLMO)" (Spinath, Stiensmeier-Pelster, Schöne, & Dickhäuser, 2002) was used. It measures learning and achievement motivation. And last but not least, "Aussagenliste zum Selbstwertgefühl (ALS)" (Schauder, 1991) was used. This measures the self-esteem in different contexts: school, family and free time activity.

Questionnaire for the parents and for the teachers. To get information about the behavior and the emotional expressions of the children, teachers and parents were asked to answer a questionnaire which measures internalizing and externalizing symptoms. The "CBCL (child behavior checklist 4–18)" (Arbeitsgruppe Deutsche Child Behavior Checklist, 1998) was used for the parental evaluation as well as for the teachers' "Teacher's Report Form (TRF)" (Döpfner, Berner, & Lehmkuhl, 1994).

4 Results

The first calculations to analyze the differences between the groups, show significant interactions between the group and the data collection ($p = .03$). Dyslexic children in integrative classes show a significant lack of scholastic self-esteem compared with the control group (from the first to the second data collection). A simple linear regression analysis shows that the scholastic self-esteem at the first data collection, predicts significantly the self-esteem at the second data collection (A $Rsquare = 234; p < .001$). The results of the scale school aversion (posthoc-analysis after ANOVA) show a significant group difference ($p = .04$). Children who are in the control group ($MV = 49.11$ and $SD = 10.07$) presented a more significant lack of school aversion than dyslexic children in integrative classes ($MV = 58.29$ and $SD = 4.35$). This result is confirmed by the Kruskal-Wallis-Test ($p = .022$) and the Welch-Test ($p = .000$). For the second data collection about school aversion, it is the same thing. Significant groups' differences ($p = .002$ and $p = .003$) between control group ($MV = 46.28$ and $SD = 9.25$) and integrative class group ($MV = 59$ and $SD = 6.66$), but also between integrative class group and dyslexic class group ($MV = 46$

and *SD* = 7.67), were calculated with the same analysis and were also confirmed with the Kruskal-Wallis-Test (*p* = .008).

At the moment there is only a descriptive view on the parental and teachers' evaluations. It refers to a difference between dyslexic children in special dyslexic classes and the control group in a way that teachers and parents of dyslexic children report more externalizing and internalizing symptoms than parents and teachers of non-dyslexic children.

5 Discussion

The first results imply the positive effects of special dyslexic classes on the subjective well-being of dyslexic children and the prevention of secondary symptoms. The further development during the future measurement points has to be analyzed.

6 Affiliation

M. Sc. Regine Eichhorn
Institution: University of Leipzig, Educational and Rehabilitation Psychology
Address: Neumarkt 9–19, 04109 Leipzig, Germany
E-mail: regine_eichhorn@hotmail.de

7 References

Arbeitsgruppe Deutsche Child Behavior Checklist. (1998). *Elternfragebogen über das Verhalten von Kindern und Jugendlichen; deutsche Bearbeitung der Child Behavior Checklist (CBCL/4–18). Einführung und Anleitung zur Handauswertung* (2. Auflage). German norm was modified by M. Döpfner, J. Blück, S. Bölte, K. Lenz, P. Melchers & K. Heim. Köln: Arbeitsgruppe Kinder-, Jugend- und Familiendiagnostik.

Betz, D., & Breuninger, H. (1998). *Teufelskreis Lernstörungen – Theoretische Grundlegung und Standardprogramm, Material für die klinische Praxis* (5. Auflage) Psychologie Verlags Union: Weinheim.

Döpfner, M., Berner, W., & Lehmkuhl, G. (1994). *Handbuch: Lehrerfragebogen über das Verhalten von Kindern und Jugendlichen. Forschungsergebnisse zur deutschen Fassung des Teacher's Report Form (TRF) der Child Behavior Checklist.* Köln: Arbeitsgruppe Kinder-, Jugend- und Familiendiagnostik.

Eichhorn, R. (2010). *Sekundärsymptome von Legasthenikern.* (Unpublished bachelor thesis). University of Leipzig, Leipzig, Germany.

Eichhorn, R. (2012). *Sekundärsymptomatik von Legasthenikern.* (Unpublished master thesis). University of Leipzig, Leipzig, Germany.

Schauder, T. (1991). Die Aussagen-Liste zum Selbstwertgefühl für Kinder und Jugendliche (ALS), In R Jäger, & F. Petermann (Hrsg.) *Treatmentorientierte Diagnostik*. Weinheim: Beltz.

Spinath, B., Stiensmeier-Pelster, J., Schöne, C., & Dickhäuser, O. (2002). *SELLMO – Skalen zur Erfassung der Lern- und Leistungsmotivation*. Göttingen: Hogrefe Verlag.

Valtin, R. (1989). Dyslexia in the german language. In P. G. Aaron & R. M. Joshi (Eds.), *Reading and Writing Disorders in Different Orthographic Systems*, (pp. 119–135). Dordrecht: Kluwer.

Wieczerkowski, W., Nickel, H., Janowski, A., Fittkau, B., & Rauer, W. (1973). *AFS – Angstfragebogen für Schüler*. Braunschweig: Georg Westermann Verlag.

Yumi Lee, Julia Strobel, & Evelin Witruk

University of Leipzig, Germany

Teachers' Knowledge about Dyslexia: A Cross-cultural Comparison Study between Germany and South-Korea

Abstract. Children with dyslexia have particular problems in learning how to read and write. Therefore they need special support. In Germany (specifically in Saxony), special dyslexia classes have been established, in which students with dyslexia are taught by qualified teachers for two years. In Korea, however, even the term of dyslexia is not well known and recognized for teachers. The aim of this study was to examine cross-cultural similarities and differences of teachers' experiences and their knowledge about children with ADHD by comparing three samples of primary school teachers from Korea and Germany (a: German special dyslexia class teachers; b: German regular class teachers; c: Korean regular class teachers). Matched samples of 45 German teachers (special dyslexia and regular classroom teachers) and 45 Korean regular classroom teachers were used for data analysis. Descriptive analysis, frequency analysis, and mean analysis were used to test 3 hypotheses. As a result, German special dyslexia class teachers have higher knowledge about dyslexia as well as students with dyslexia. However, regular class teachers (both in Korea and Germany) are not sufficiently prepared to teach students with dyslexia and to support these individually. Compared to German sample (both special and regular class teachers), Korean teachers have significantly lower knowledge about dyslexia. Thus, (a) awareness, (b) study/information, and (c) additional training (at university, in-service) of dyslexia are urgently needed for Korean teachers and conducted, so that students with dyslexia could receive suitable attention and education in school.

Keywords: special dyslexia class, regular class, learning problems, cross-cultural study.

1 Introduction

Students with dyslexia have severe academic problems due to their reading and writing difficulties. Therefore they need special support. In Germany, special dyslexia classes have been established in Saxony so that students with dyslexia can be provided lessons taught by qualified teachers for two years. In the course of the debate on inclusion, German education policy suggests to abandon these special dyslexia classes, which would have far-reaching consequences. In Korea, however, even the term 'dyslexia' is not well known, thus fewer studies have been conducted with regards to the topic of dyslexia, and none of these studies focus on teachers' knowledge about this affliction. Since both German and Korean children with dyslexia spend five days a week at school, teachers are important people to notice dyslexia-related problems, and to refer these children to professionals for an accurate diagnosis and treatment. Therefore, it is essential for both German and Korean teachers to have accurate information and knowledge about dyslexia in order to help children who potentially have reading and writing problems, as well as to provide correct advice to their parents. The purpose of this study was to investigate cross-cultural similarities and differences between Korean and German teachers with regards to their various experiences and knowledge regarding students with dyslexia within their own culture as well as across cultures.

2 Research questions and hypotheses

The following research questions were addressed by the study. Specific hypotheses were then made for each of the goals of this study.

- Research question 1 (Germany vs. Korea)
  Are there significant differences between German and Korean teachers in their knowledge about dyslexia?
- Hypothesis 1
  German teachers would be more knowledgeable about students with dyslexia than Korean teachers.
- Research question 2 (special dyslexic class vs. regular class)
  Are there significant differences between teachers in special dyslexia classes and teachers in regular classes?
- Hypothesis 2
  German special dyslexia class teachers would have higher level of knowledge regarding students with dyslexia in comparison to Korean and German regular class teachers.
- Research question 3 (experienced teachers vs. less experienced teachers)
  Are there significant differences between teachers with more experiences of dyslexia and those with little-to-no such experiences?
- Hypothesis 3
  Teachers with more experience of dyslexia would have significantly greater knowledge of students with dyslexia in comparison to those who have little-to-no such professional experience in both countries (H3a: Germany; H3b: Korea).

3 Method

3.1 Sample

Matched samples of 45 German teachers (25 qualified dyslexia teachers and 20 regular classroom teachers) and 45 Korean regular classroom teachers were used for data analysis.

3.2 Survey instruments

Based on previous research, a German version of the survey instrument was developed by second and third author of this study, which was divided into three sections with a cover letter: Section A (Experiences), Section B (Knowledge), Section C (Personal Details). For a Korean version of the survey instrument, translation/back-translation procedure and item review were conducted by professionals in Korea (the first author, a PhD student of University Bonn, and a professor of Korean National University of Education) in order to confirm the equivalence of the survey instrument between Germany and Korea. The actual study was conducted in Germany from February to March 2014 and in Korea from July to August 2014.

3.3 Data analysis

Descriptive analyze, frequency analyze, and mean analyze (t-test, ANOVA) were used to test three hypotheses.

4 Results

4.1 Hypothesis 1 testing

German teachers ($M = 14.37$, $SD = 2.22$) had higher knowledge about dyslexia (children with dyslexia) than Korean teachers ($M = 9.97$, $SD = 4.43$), $t(64.860) = 5.942$, $p < .001$, $d = 1.25$ (H1 supported) (see *Figure* 1).

4.2 Hypothesis 2 testing

German special dyslexia class teachers ($M = 15.32$, $SD = 1.91$) have higher level of knowledge regarding students with dyslexia in comparison to Korean ($M = 9.97$, $SD = 4.43$) and German regular class teachers. ($M = 13.21$, $SD = 2.06$), $F(2, 87) = 20.394$, $p < .001$, $d = 1.25$ (H2 supported) (see *Figure* 2).

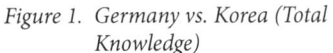

Figure 1. Germany vs. Korea (Total Knowledge)

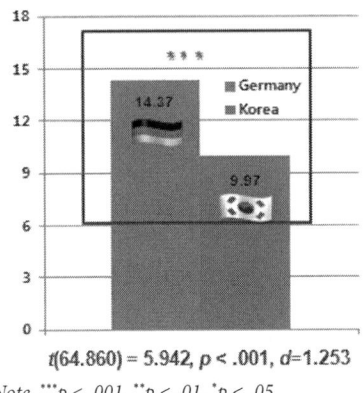

$t(64.860) = 5.942, p < .001, d=1.253$

Note. $^{***}p < .001$, $^{**}p < .01$, $^{*}p < .05$

Figure 2. Special Dyslexic vs. Regular Class (Knowledge by Class-Type)

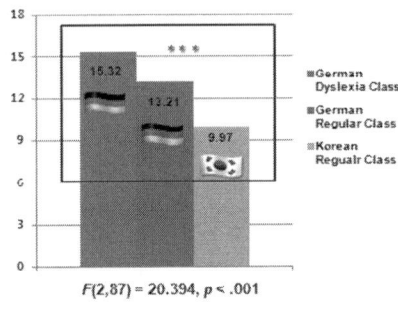

$F(2,87) = 20.394, p < .001$

Note. $^{***}p < .001$, $^{**}p < .01$, $^{*}p < .05$

4.3 Hypothesis 3 testing

In Germany, teachers with high experience of dyslexia ($M = 15.21$, $SD = 1.47$) had higher knowledge of students with dyslexia in comparison to those who have low experience ($M = 14.44$, $SD = 2.40$) and middle experience ($M = 13.21$, $SD = 2.48$). However, the differences between groups were not significantly different: $F (2, 42) = 1.901$, $p = .162$ (H3a rejected) (see Figure 3). In Korea, no teachers have higher level of knowledge ($M = 0.00$, $SD = 0.00$), about dyslexia (children with dyslexia). No significant difference was found between teachers with middle ($M = 10.00$, $SD = 4.34$), and low experience of dyslexia ($M = 9.96$, $SD = 4.55$), $F (1, 43) = .001$, $p = .981$ (H3b rejected) (see Figure 4).

Figure 3. Experiences and Knowledge (German sample)

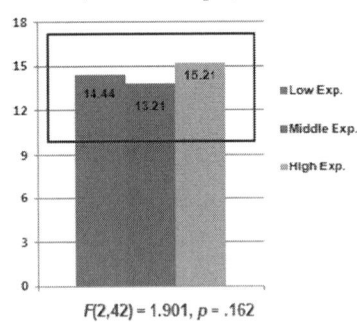

$F(2,42) = 1.901, p = .162$

Note. $^{***}p < .001$, $^{**}p < .01$, $^{*}p < .05$

Figure 4. Experiences and Knowledge (Korean sample)

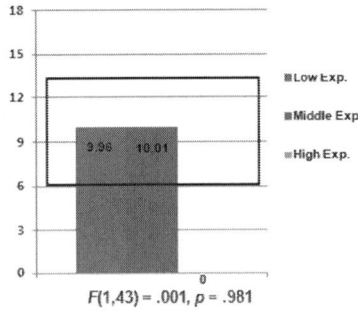

$F(1,43) = .001, p = .981$

Note. $^{***}p < .001$, $^{**}p < .01$, $^{*}p < .05$

Further analysis was conducted to more carefully assess teachers' knowledge about dyslexia (children with dyslexia) as follows: (a) special versus regular class teachers in Germany: German special dyslexic class teachers ($M = 15.32$, $SD = 1.90$) had significantly higher knowledge about dyslexia (children with dyslexia) in comparison to German regular class teachers ($M = 13.21$, $SD = 2.06$), $t(43) = 3.568$, $p = .001$, $d = 1.06$ (see *Figure* 5); (b) regular class between countries: German regular class teachers ($M = 13.21$, $SD = 2.06$) had significantly higher knowledge about dyslexia (children with dyslexia) in comparison to Korean regular class teachers ($M = 9.97$, $SD = 4.43$), $t(62.793) = 3.992$, $p = .003$, $d = 0.93$ (see *Figure* 6).

Figure 5. Special vs. Regular class within Germany

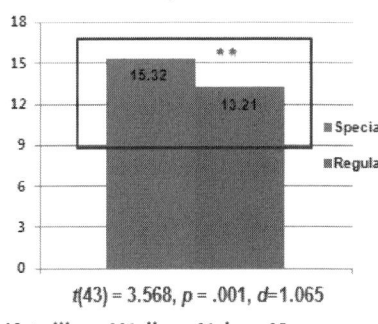

$t(43) = 3.568$, $p = .001$, $d=1.065$

Note. ***$p < .001$, **$p < .01$, *$p < .05$

Figure 6. Germany vs. Korean Teachers in Regular class

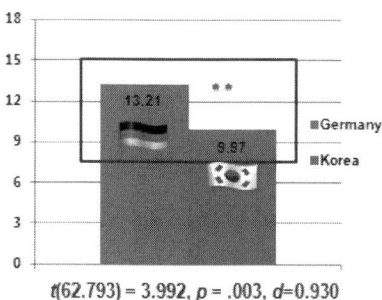

$t(62.793) = 3.992$, $p = .003$, $d=0.930$

Note. ***$p < .001$, **$p < .01$, *$p < .05$

5 Discussion

German special dyslexia classroom teachers have higher knowledge about dyslexia as well as about students with dyslexia. However, regular classroom teachers (both in Korea and Germany) are not sufficiently prepared to teach students with dyslexia. The conclusion taken from the results is that regular primary school teachers in both countries are not sufficiently prepared to teach children with dyslexia and to support them individually. In addition, this argues against the abolition of German dyslexia classes, as long as the conditions for an inclusive education are not created yet. Compared to German sample (both special and regular classroom teachers), Korean teacher have significantly lower knowledge about dyslexia. Thus, (a) awareness, (b) study/information, and (c) additional training (at university, in-service) of dyslexia are urgently needed for Korean teachers to be investigated and conducted, so that students with dyslexia could receive suitable attention and education in the classroom.

6 Affiliations

Dr. Yumi Lee
Institution: University of Leipzig, Educational and Rehabilitation Psychology
Address: Neumarkt 9–19, 04109 Leipzig, Germany
E-mail: yumi.lee@uni-leipzig.de

M.Sc. Julia Strobel
Institution: University of Leipzig, Educational and Rehabilitation Psychology
Address: Neumarkt 9–19, 04109 Leipzig, Germany
E-mail: psy12mgt@studserv.uni-leipzig.de

Prof. Dr. Evelin Witruk
Institution: University of Leipzig, Educational and Rehabilitation Psychology
Address: Neumarkt 9–19, 04109 Leipzig, Germany
E-mail: witruk@uni-leipzig.de

7 References

Dilling, H., Mombour, W. & Schmidt, M. H. (Eds.). (2010). *Internationale Klassifikation psychischer Stoerungen: ICD-10.* (7. Ed.). Bern: Huber.

Esser, G. & Wyschkon, A. (2011). Umschriebene Entwicklungsstoerungen. In G. Esser (Ed.), *Lehrbuch der klinischen Psychologie und Psychotherapie des Kindes- und Jugendalters* (pp. 159–181). Stuttgart: Thieme.

Hudson, R. F., High, L., & Al Otaiba, S. (2007). Dyslexia and the brain: what does current research tell us? *The Reading Teacher, 60,* 506–515.

Strobel, J. (2014). Erfahrungen, *Wissen und Einstellungen von Lehrkraeften zu Schuelern und Schuelerinnen mit Lese-Rechtschreibschwaeche: Regel-Grundschullehrer und LRS-Lehrer im Vergleich.* (Unpublished master thesis). University of Leipzig, Leipzig, Germany.

Washburn, E. K., Joshi, R. M., & Binks-Cantrell, E. S. (2011). Are preservice teachers prepared to teach struggling readers? *Annals of Dyslexia, 61,* 21–43.

Warnke, A., Hemminger, U. & Plume, E. (Eds.). (2004). *Lese-Rechtschreibstörungen. Leitfaden Kinder- und Jugendpsychotherapie.* Göttingen: Hogrefe.

Samudra Senarath

University of Colombo, Sri Lanka

Teachers' Knowledge about Dyslexia in Sri Lanka

Abstract. For many years scientists have conducted studies on reading disability where some children have great difficulty in learning to read, write, spell and pronounce the words, despite normal intelligence. These difficulties are identified under the concept of dyslexia. Researchers have proved that dyslexic symptoms shows poor letter-sound knowledge and phoneme awareness, as well as problem with coping. Further studies proved that dyslexic children possesses anxiety, psycho somatoform complaints, and less self-esteem however teachers' awareness of these problems are less. Research studies show that there are misconceptions, lack of awareness and knowledge among educators concerning dyslexia. This lack of awareness among teachers leads to a poor understanding of the needs of children which results in anxiety. The teacher's self effectiveness, mental status towards them plays a significant role. Teachers' who acquire knowledge and training, may have better chances to identify children with dyslexia, meet their diverse needs, and may mediate the adjustment of those needs. In Sri Lanka this type of research is lacking. But in school settings, most of the teachers have recognized that some students find difficulty in reading and writing even though the teachers' knowledge about the dyslexic children is less. So the objectives of this study were to identify teachers' knowledge towards dyslexic children, examine teachers' effectiveness and investigate teachers' emotions when they work with dyslexic children. The survey method and purposive sampling techniques were used for this study. Accordingly, one hundred teachers were selected from 15 schools by school principles' nomination including primary, secondary and special education teachers. Standardized test were carried out on subjective and objective knowledge on dyslexia, teachers mental statues and self-effectiveness were used in the study. The findings on the teacher participation for the special training program related to the concept of dyslexia showed that, 98 % of special education teachers', 32 % of primary teachers and 1 % of secondary teachers participated for the training. Moreover, Special education teachers have been trained well in contrast to the other teacher groups. Special education and primary teachers had sufficient knowledge concerning dyslexia characteristics but secondary teachers had insufficient knowledge and misunderstood the cause of dyslexia. The results revealed that special education teachers had positive emotions and self effectiveness compared to the other teacher groups. This study reveals that there is a need of handing out more teacher training and facilities; learning aids, teacher student ratio concerning the dyslexic children.

Keywords: dyslexia, special education, teacher knowledge.

1 Introduction

During the last century hundreds of scientists searched the specific sources of reading disability which is some children have great difficulty in learning to read and write despite normal intelligence, and in the absence of any other factor which could conceivably impede their learning, a sight or hearing problem. These difficulties are identified under concept of dyslexia. The definition of dyslexia has been debated for long time and the dyslexia literally means difficulty in reading, writing, spelling, and pronouncing words (Snowling, Duff, Petrou, & Schiffeldrin, 2011). Although many researchers have tried to define dyslexia by applying various theories, there are three prominent theories used in different studies to explain dyslexia such as the phonological theory, the magnocellular theory and the cerebellar theory. World Health Organization (WHO) also has introduced a restricted developmental disorder in the acquisition of reading often connected with a disorder in acquisition of writing (more information see ICD 10- F81). These disorders are usually contrasting over the better, normal or over-averaged intelligence.

The prevalence rates of dyslexia in the world vary from 1 % in Scandinavian countries, 3 %-5 % in Germany, 8 %-10 % percent in UK and USA. This leads to the question of the cultural/language impact on the development of dyslexia in a specific language /cultural environment. The relation of boys to girls is about 4:1. Fifteen to twenty percent of the population copes with different symptoms of dyslexia, such as poor spelling, reading fluency and difficulties in expressing themselves (Shaywitz, 2003).

Characteristic features of dyslexia are difficulties in phonological awareness, verbal memory and verbal processing speed. Researchers pointed out that some children may make good progress from early word reading development, but show signs of difficulties with literacy later in development. There is some evidence that such children may be those with good oral language skills, who nonetheless show early delays with spelling acquisition and subsequent reading fluency difficulties (Snowling et al., 2011). Researchers have shown that dyslexic symptoms of a preschool child as delayed or problematic speech, poor expressive language, poor rhyming skills, little interest/difficulty in learning letters and in early school child symptoms shown are, poor letter-sound knowledge, poor phoneme awareness, poor word skills and problem with coping. Research studies proved that dyslexic child presents anxiety, psycho somatoform complaints, coping problems, and less self-esteem but teachers awareness of this problem is less (IDA 2004; Witruk, 2010).

With appropriate teaching and support, dyslexic children do make progress but reading and writing are always likely to be effortful for them. Some research in this field highlighted that many teachers are aware of this problem, but lack of time, and being overloaded with daily school routines and responsibilities, prevent them from helping and assessing these children. On the other hand some countries have conducted popular surveys that highlighted the alarming rate of increase in dyslexia and have started a national initiative program called Dyslexia friendly schools. In contrast, some countries, for many years' teachers have been concerned about students who appear normal, intelligent, and healthy, but struggle with reading and learning to read and write.

It is highly likely that, those teachers' abilities in dealing with different forms of learning difficulties of those children will be dependent on these teachers' knowledge and attitudes towards those children's difficulties. Research studies show that there are misconceptions, lack of awareness and knowledge among educators concerning dyslexia Roper 2010; Reid, 2005; Wadlington & Wadlington, 2005). This lack of awareness among teachers does not help to understand the needs of children, and this can result in anxiety (Reid, 2005). According to Wadlington and Wadlington (2005), dyslexic student's self-esteem can be fragile, and the teacher's attitude towards them plays a significant role during their studies. Educators, who acquire knowledge and training, may have better chances to identify children who may be coping with dyslexia, meet their diverse needs, and may mediate the adjustment of those needs.

Review of relevant literature reveals that dyslexic children are having reading and writing difficulties. In Sri Lanka although this type of research is lacking, in school settings most of the teachers have recognized students that have difficulty in reading and writing. But the teachers' knowledge about the dyslexic children is less. So the purposes of this study were to identify teachers' knowledge about the dyslexic children, examine teachers' effectiveness, and investigate teachers' emotions level when they work with dyslexic children in teaching-learning and to examine teachers' job satisfaction.

2 Literature

There are numerous studies done for educators awareness and understanding of learning disabilities in general (Kirby, Davies, & Bryant, 2005; Wight & Chapparo, 2008). However, there is a lack of research done for the awareness of teachers concerning the knowledge about the children with dyslexia. One of the researchers that do concentrate on dyslexia and teachers understanding of dyslexia is that of Wadlington and Wadlington (2005), "What Educators Really Believe about Dyslexia".

Further Wadlington and Wadlington (2005) conducted a study were they compared the perception and knowledge of dyslexia among 250 participants at a southern regional university and faculty members in US. They developed a 30-item survey, which they named it as Dyslexia Belief Index (DBI). In their study Wadlington and Wadlington (2005) found that the educators had insufficient knowledge concerning dyslexia and most of them misunderstood the case of dyslexia. According to their findings clear misconception was found in most of the educators, Wadlington & Wadlington (2005) determined that there is a need of handing out more information and trainings to the educators concerning this reading specific learning disability called dyslexia.

Another research study done by Ashburn and Snow (2011) relating on the importance of teacher's awareness and knowledge concerning children coping with dyslexia and intervention program, with its aim to give the teachers a better understanding about dyslexia. The intervention program was created for elementary, secondary and special education teachers. Results were proved that prior to the intervention teachers do have misconceptions about dyslexia as well as less knowledge in this area.

The study of Washburn, Joshi and Binks-Cantrell (2011), is another study with a purpose of identifying teachers' knowledge about different language concepts and dyslexia. The participants in this study were elementary school teachers. The participants were collected form two data groups. Group one consisted of 99 participants and was from 10 different schools of the district in a Midwestern state in the United States, group two, consisted of 86 participants from an urban school district in Southwest United States. In this study, the researchers found that the teachers carried a common misconception of what dyslexia is. They were confusing dyslexia with a "visual processing deficit rather than phonological processing deficit" (Washburn et al., 2011). Lee (2009) conducted a research in China for teachers' perceptions of children with dyslexia and researcher obtained the perceptions of students with dyslexia from ten Chinese and English Language teachers. The teachers were working in a local primary school, in Hong Kong. Except qualitative interviews, Lee also used the DBI tool to measure the perception of teachers concerning dyslexia. Using the DBI tool, it was "found that all language teachers held a number of misconceptions about dyslexia, but they had a better understanding of teaching strategies for dyslexic students" (p. 1). However, from the qualitative interviews he found out "the collaboration and cooperation between different school stakeholders was not well-developed and exam-oriented learning atmosphere restricted how teachers perceived and helped dyslexia" Lee (2009, p. 1). As stated previously, there is a lack of research studies concerning the knowledge of dyslexia among teachers. It is as well assumed that there is a lack of research on dyslexia's awareness and knowledge among educators in Sri Lanka.

3 Method

3.1 Sample

The survey method and purposive sampling techniques were employed for the present study. Accordingly, one hundred graduated teachers were selected from 15 schools by school principles' nomination including primary, secondary, and special education teachers. Age categories of teachers were: 25 yrs–35 yrs (24 %) and 36 yrs–45 yrs (22 %) primary teachers; 23 yrs–35 yrs (32 %) and 36 yrs–45 yrs (4 %) secondary teachers and 24 yrs–35 yrs (4 %) and 36 yrs–45 yrs (14 %) special education teachers respectively.

3.2 Measurement tools

Questionnaire: The standardized questionnaires were employed for the present study. These are (a) Dyslexia subjective knowledge (Smith, Fabriger, Macdougal, & Wiesenthal, 2008; α = .89), and (b) Dyslexia objective knowledge (Wadlington &Wadlington, 2005; α = .54). The standardised questionnaires were translated by three academics in p sychology field. They considered whether it was culturally sound for the teachers' in Sri Lanka. The reliability of the instrument presents in this study: subjective knowledge (α = .86) and objective knowledge (α = .53).

Data analyzing methods: Quantitative data analyses were used (e.g., SPSS 16.0). The data gathered from three teacher groups and teachers subjective knowledge on dyslexia was analyzed using mean and standard deviation, background information, objectives knowledge, teachers' self-effectives and job satisfaction were analyzed using percentages, and teachers' emotions, compared by paired t-test.

4 Results

One hundred primary, secondary and special education graduated teachers responded to a survey expressing their subjective and objective knowledge on dyslexia, teachers self effectiveness, and teachers mental statues (emotions) when working with dyslexic children. The results of this study showed 63 % of teachers in the sample have less than five years' experience in teaching. Thirty seven percent teachers presented 6 to 20 years teaching experiences.

The results on the nature of teacher training of the sample shows that, 70 % primary teachers have received primary teacher training and 30 % primary teachers received special education training 30 %. Special education teachers mostly received special education training and it was 92.3 % of the sample. In addition to that, 7.7 % of them have obtained primary teacher training too. But 23 % secondary teachers have received primary training and 77 % have received subject teacher training. The results on the teacher participation of special training program on dyslexia shows, 98 % participation of special education teachers, 32 % participation of primary teachers and 1 % participation of secondary teachers. Results on the training of educational psychology and psychotherapy knowledge related to dyslexia show 100 % for special education teachers but 1 % for other teacher groups. Also special teachers are well trained in contrast to the other teacher groups.

The result showed that the teachers' subjective knowledge level on dyslexia, in primary teacher group was (M = 2.60, SD = 1.43), special education teachers was (M = 5.47, SD = .62) and secondary teacher group was (M = 1.98, SD = 1.41). According to these results special education teachers have high knowledge on dyslexic theme when compared to the other teacher groups.

Teachers' knowledge on dyslexic characteristics showed that most of the teachers have heard about dyslexia, but there is a clear misconception of what dyslexia is? According to the *Figures* 1 and 2, teachers' knowledge about dyslexic children on the characteristic "very often they mix up letters when reading and writing" was 46.9 % correct responses by primary teachers and 24 % correct responses by secondary teachers. Incorrect responses for the same were 53.1 % and 76 % primary and secondary teachers respectively. When comparing the teachers' knowledge, special education teachers showed highest awareness about the dyslexic characters. Moreover, they showed high awareness on "unable to pronounce words correctly" compared to other teacher groups (see *Figure* 2).

Figure 1. Dyslexic Characteristics: Mixed   *Figure 2. Dyslexic Characteristics:*
up Letters When Reading and   *Unable to Pronounce Words*
Writing   *Correctly*

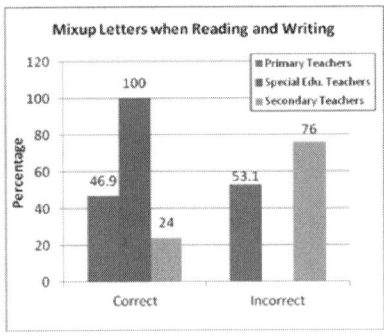
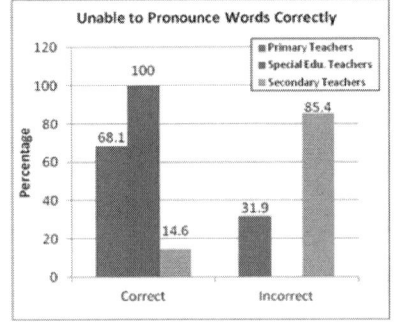

On teachers' job satisfaction under the character "satisfaction of number of children in the classroom", 56 % primary teachers stated that they are unsatisfied. It was 13.7 % among special education teachers and 23 % among secondary teachers. In contrast, when considering satisfaction, 44 % primary teachers, 1.2 % secondary teachers and 86.3 % special education group was satisfied. Satisfaction on teacher salary differed by teacher groups. 42.8 % of primary teachers, 70 % of secondary teachers and 100 % special teachers were unsatisfied. Accordingly teachers are unhappy about their salary scales while they are working on reading and writing difficulties with children. Teachers self -effectiveness when working with dyslexic children measured as "able to help with suitable exercises, activities and teaching methods" was shown as 78 % for primary teachers, 97 % for special education teachers and 38 % for secondary teachers through teaching -learning process. Primary and special education teachers have positive self-effectiveness in contrast to the secondary teacher category.

Figure 3. Teachers' Mental Status

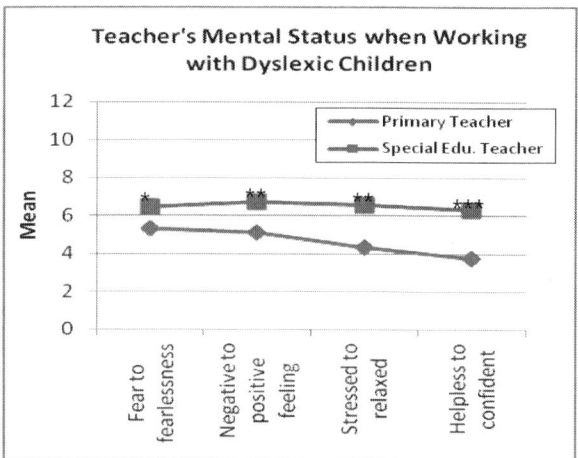

The results revealed that teachers' mental status under emotions represented negative versus to positive level when working with dyslexic children at the classroom. According to *Figure* 3, fear to fearlessness of primary teachers were (M = 5.28, SD = 1.08) and (M = 6.46, SD = .91) special education teachers (p <.05). Results on helpless to confident showed (M = 3.76, SD = 1.42) for primary teachers and (M = 6.30, SD = 1.23), (p < .001) for special education teachers. According to *Figure* 3, special educational teachers have positive mental status in contrast to the primary teacher group. Above teacher groups have significantly different emotions through teaching leaning process.

5 Discussion

Participation of special education teachers was more in the training program and they presented higher knowledge about dyslexia. Teachers' knowledge on dyslexic characteristics showed that most of the teachers have heard about dyslexia, but there were clear misconceptions of what dyslexia is? Primary teaches also had certain level of knowledge but secondary teacher group had less knowledge. Most of the teachers had misunderstood the concept of dyslexia. There is a significant difference between primary and special education teachers' emotions when they are working with reading and writing problem children. Based on this research study teachers need more training on dyslexia, more facilities, teaching -learning aids, collaborative works and additional assistant teachers for primary grades etc. Necessity of increasing teachers' job satisfaction and suitable teacher student ratio for the primary grades is also brought forward.

6 Affiliation

Dr. Samudra Senarath
Institution/ Address:
Department of Educational Psychology,
University of Colombo, Sri Lanka
E-mail: pgrsamudra@yahoo.com

7 References

Ashburn, D, L., & Snow, B, K. (2011). *Dyslexia: Awareness and intervention in the classroom*. Retrieved from http://csus-dispace.calstate.edu/bitstream/1.

International Dyslexia Association. (2004). *Social and emotional problems related to dyslexia*. Retrieved from http://www.interdys.org/ewebeditpro5/upload/Social and_Emotion_ Problems_Related_to_Dyslexia.pdf.

Kirby, A., Davies, R., & Bryant, A. (2005). Do teachers know more about specific learning difficulties than general practitioners? *British Journal of Special Education, 32*, 122–126.

Lee, F. (2009). *Teachers' perceptions of students with dyslexia in a local primary school*. (Unpublished doctoral dissertation). University of Hong Kong, Hong Kong, China.

Ramus, F. (2003). Theories of developmental dyslexia: insights from a multiple case study of dyslexic adults. *Brain, 126*(4), 841–865.

Reid, G. (2005). *Dyslexia and inclusion, classroom approaches for assessment, teaching and learning.* London: David Fulton Publisher Ltd.

Roper, G. (2010, December). *New poll reveals dangerous confusion about children with Learning disabilities.* The Examiner; International Dyslexia Association.

Shaywitz, S. E. (2003). *Overcoming dyslexia: A new and complete science-based program for reading problems at any level.* New York: Alfred A. Knopf.

Smith, S. M., Fabrigar, L. R., Macdougall, B. L., & Wiesenthal, N. L. (2008). The role of amount, cognitive elaboration, and structural consistency of attitude-relevant knowledge in the formation of attitude certainty. *European Journal of Social Psychology, 38* (2), 280–295.

Snowling, M., Duff, F., Petrou, A., & Schiffeldrin, J. (2011). Identification of children at risk of dyslexia: The validity of teacher judgements using 'Phonic Phases'. *Journal of Research in Reading, 34*, 157–170.

Wadlington, E. K., & Wadlington, P. L. (2005). What educators really believe about dyslexia? *Reading Improvement, 42*, 16–33.

Washburn, E. K., Joshi, R. M., & Binks-Cantrell, E. S. (2011). Teacher knowledge of basic language concepts and dyslexia. *Dyslexia (Chichester, England), 17*(2), 165–83.

World Health Organization. (2011). *Annual World Report on Disability.* Retrieved from: http://www.who.int/disabilities/world_report/2011/en/index.html.

Wight, M., & Chapparo, C. (2008). Social competence and learning difficulties: Teacher perceptions. *Australian Occupational Therapy Journal, 55*, 256–265.

Witruk, E. (2010). Dyslexia-assessment and treatment. In E. Witruk, D. Riha, A. Teichert, N. Hasse, & M. Stueck (Eds.), *Learning, Adjustment, and Stress Disorders* (pp. 41–57). Frankfurt: Peter Lang.

Buddhiprabha D. D. Pathirana

University of Peradeniya, Sri Lanka

Rainbow Forever: Recommendations and Suggestions for Potential Psychosocial Interventions to Sri Lankan Children with Dyslexia

Abstract. Specific learning disabilities, such as dyslexia, remain a concern for the children, families involved, and educationists/ education systems. Children with dyslexia may have varied psychosocial issues, due to the necessity of being expected to and their desperate attempts to fit into a so called 'normal reading/writing class'. Hence, left unaddressed, dyslexia may lead to frustration, low self-confidence, and poor self-esteem which may substantially increase the risk of developing psychological and emotional problems within these children. Even though scientific research on dyslexia during the past few decades has revealed a great deal about its nature, aetiology and assessment, in the Sri Lanka milieu extremely few measures have been carried out in this area. As a result, only a handful of diagnostic and remedial programs are available for Sri Lankan children with dyslexia (SCwD), with majority of them being provided by professionals in private, fee levying contexts. While recognizing the importance of providing universal early detection and remedial programs to all SCwD, the present study explores the psychosocial services available to them using semi-structured interviews (with experts, teachers and SCwD). The study also provides recommendations to potential changed agents (policy makers, primary school teachers, teacher trainers, clinicians, and parents of SCwD) of the micro, macro and meso-levels to improve and develop the situation of SCwD.

Keywords: Sri Lanka, dyslexia, psychosocial interventions.

1 Introduction

Scientific research on dyslexia has proliferated over the past five decades. As a result, plethora of research exits pertaining to its nature, aetiology and assessment in the globe. Against this backdrop, it should be possible for teachers, psychologists and other child mental health professionals/workers to recognise the signs which suggest that a child is at risk of reading failure. Such early identification should allow remedial actions to be executed before a sense of downward snow balling spiral of underachievement, low self-esteem and poor motivation is established within the children with dyslexia, setting them on destructive developmental pathways.

Literature convey, if unaddressed; dyslexia may lead to psychosocial issues such as frustration, low self-confidence, and poor self-esteem (Glazzard, 2012) which may substantially increase the risk of developing psychological and emotional problems within these children. Studies also depict that students with dyslexia have negative attitude toward reading, reading attitude and little motivation to read in comparison to students without dyslexia (Mihandoost, Elias, Sharifah, & Mahmud, 2011).

Over the past decade, the Sri Lankan government has taken significant strides to improve the quality of the special education in Sri Lanka (UNICEF, 2003). Among them are 'the nation-wide community-based rehabilitation programs', inclusive education, pre/school teacher awareness/ training, training of trainers, provisions of essential resources, and setting up of special educational units within the regular schools (UNICEF, 2003). In spite of providing all these facilities/provisions; developing remedial measures for SCwD seemed to be still in its infancy due to following reasons. First, a standardized tool to screen dyslexia remains absent to date. Second, less than a handful of diagnostic and remedial programs are available for SCwD, with majority of them being provided in Colombo (Capital of Sri Lanka); by professionals in private, fee levying contexts.

The purpose of the present study was to explore the psychosocial interventions available to ScwD. To achieve this purpose, varied stakeholders (those providing and receiving psychosocial services pertaining to dyslexia) in Sri Lanka were interviewed using open-ended, semi structured interview format.

2 Method

In order to optimize the results of the study, purposive sampling was used. Participants were selected from varied ecological tiers (i.e., Micro, Meso, Exo and Macro systems, based on Urie Bronfenbrenner's ecological theory); providing and receiving psychosocial services pertaining to dyslexia in the Sri Lankan milieu. When selecting participants, the researcher considered the diversity of their experiences pertaining to psychosocial interventions of dyslexia. The participants were experts conducting research or providing psychosocial services to SCwD ($n = 4$), special education teachers ($n = 3$), Children with Dyslexia ($n = 2$; Boy = 12 years; Girl = 14 years) and regular teachers ($n = 50$) who teach non dyslexic children.

3 Results

3.1 Experts

Interviews with the experts conveyed that few measures have been taken to provide psychosocial intervention for SCwD. Though special education courses are available in national universities, specific knowledge on dyslexia in terms of interventions or psychosocial support is not provided to special education teachers, parents or children with dyslexia. Expert interviews also conveyed that a handful of national/ international schools in Sri Lanka have taken initiatives to train their special education teachers through local/.foreign experts who have been trained abroad. Expert interviews further conveyed that due to absence of screening tools or interventions; SCwD have been marginalized and their right to education violated (e.g., being requested to leave regular schools, being reprimanded and punished for negligence). Girls with dyslexia seem to experience more segregation than boy in the traditional Sri Lankan society, with more girls dropping out of school in the secondary educational classes. According to experts there has been an emerging interest to conduct research on dyslexia, with one ongoing research to explore its prevalence. However, very few diagnostic and remedial programs are available for SCwD, with majority of them being provided by professionals in private, fee levying contexts.

3.2 Teachers

Group discussion carried out with regular teachers, teaching non-disabled children conveyed that they did not have the knowledge and the competency to identify children with dyslexia, provide remedial educational measures/psychosocial/psycho-educational interventions for SCwD or their parents. Interviews with special education teachers ($n = 2$) conveyed SCwD experience many difficulties, psychosocial as well as educational. According to them, special provisions are not available to SCwD within the Sri Lankan educational system (e.g., provision of extra time during national examinations, even when requested). However, this privilege is granted to Sri Lankan children with disabilities (i.e., visually/auditory impaired) other than dyslexia. As a result, special education teachers perceived SCwD to be more marginalized in comparison to children with physical disabilities. Interviews also conveyed even the special education teacher seem to perceive dyslexia as an 'invisible disability' resulting SCwD receiving less attention than children with so called other disabilities.

3.3 Children

Interviews with SCwD ($n = 2$) attending a school for children with special needs conveyed that they experience many psychosocial issues such as marginalization from the community and their non-disabled peers. The SCwD mentioned that in their former schools (of which the majority of the children were without any form of apparent disability); they experienced corporal punishment, emotional abuse and neglect due to the fact that teachers who work with non-disabled children attributed their inability to negligence, 'stupidity' and inefficiency. The children also mentioned even though they presently attend a special school, they are still being marginalized and bullied by their non-disable peers and neighbours simply due to the fact that they 'go to a special school'. This seemed to create anger, distress, frustration and sadness within the child participants.

4 Discussion

Presently, psychosocial interventions provided to SCwD appear to be almost non-existent in the Sri Lankan milieu. As a result, SCwD seem to be burdened with psychosocial problems in addition to the neurological and educational challenges they experience due to dyslexia. Hence, there is an urgent need to explore/develop culturally appropriate psychosocial interventions for SCwD and their families. Hence, in order to provide an effective sustainable change and remove barriers to academic and personal achievement that SCwD experience; the present article recommends that significant individuals and service providing systems (e.g., schools, special educational centers) must use an ecological and sociocultural approach (Bronfenbrenner, 1970). When developing psychosocial interventions; the impact of the interactions between home, school, sociocultural environments, as well as the impact of the macro level ecological systems have on SCwD need to be acknowledged.

The article specifically emphasises the significance of macro level interventions involving policy level changes that shifts the attention away from the SCwD. It also proposes to create a collaborative intervention that envelops individuals, families, and communities. The article is of the opinion that such an intervention would shift the focus from the child to the community, challenging the present existing belief pertaining to dyslexia (i.e., dyslexia being a factor which occurs due to the negligence of the child). As a result, SCwD are being perceived as there is 'something wrong with him/her', and 'he/she need to be fixed to suit the society'. The study also recommends the policy makers to commission for a national level research studies to identify nature, prevalence, and etiology of dyslexia; develop and standardized a screening tool to assess dyslexia; identify and develop culturally feasible interventions; develop resource materials to create awareness and mobilize interested parties (government, non-governmental, community stake holders, media etc.).

The present study recognizes regular teachers as the most important human resource for promoting awareness and psychosocial interventions to ScwD. However, it is aware of the fact that regular classroom teacher has little time, knowledge, competencies, and support. Therefore, the article believes that a substantial effort should be made to raise the awareness regarding dyslexia among regular classroom teachers thorough promoting culturally relevant teacher training programs/psychosocial awareness creation workshops. In addition, curriculum revisions seem an urgent need to create awareness pertaining to dyslexia within the pre-service teachers or teacher trainees. Though special education is addressed to some extent in the pre-service teacher training programs the article suggests that it should be carried out more extensively to create an increased sensitivity, knowledge and skills among the pre-service teachers. The article also recommends the educators and teacher trainers to develop a comprehensive training package on dyslexia for teacher trainees as well as teachers.

The article recognizes the crucial role of the special education teacher as a convenor, facilitator, and changed agent for the entire school; when developing/providing psychosocial interventions for SCwD. The article recommends development of resource materials, training programs, and on-site support for special education teachers to increase their skills, competencies and attitudes pertaining to dyslexia.

As the recipients who have been endowed with the 'invisible' disability' the article feels that SCwD may be experiencing extreme difficulties being subjected to bullying, corporal punishment and emotional neglect/abuse. Hence the article recommends empowerment programs to promote psychosocial wellbeing of SCwD (e.g., activities to increase self-esteem, assertiveness, confidence, reading motivation, and creating support networks), and developing resource materials to create awareness of their non-dyslexic peers (stories, books, leaflets, book marks, stickers). The article is of the opinion that this would help not only in sharing the teaching/ learning responsibility but also in promoting a non-discriminative attitudes and potential formation of social relationship in the school environment.

The article acknowledges the stress and turmoil that the parents of children with dyslexia (PoCwD) experience in a country which does not have an organized system to address it and perceives dyslexia as a phenomenon which occurs due to the negligence of the child concerned. Therefore, the article recommends providing psychosocial support to PoCwD through awareness and training, continuous support for parents through parent support groups, development and ensuring the availability of resource materials (awareness booklets, poster, book marks, etc.).

In the Sri Lankan milieu, psychosocial interventions available to SCwD are almost non-existent which calls for appropriate measures, urgently; to provide better futures for them.

5 Affiliation

Dr. Buddhiprabha D. D. Pathirana
Institution/ Address:
Department of Philosophy & Psychology,
University of Peradeniya, Peradeniya, Sri Lanka
E-mail: buddhiprabhap@pdn.ac.lk; buddhiprabha2001@yahoo.com

6 References

Bonfenbrenner, U. (1970). *Two world's of childhood: US and USSR*. London: George & Unwin Ltd.

Hulme, C., & Snowling, M. J. (2011). Children's reading comprehension difficulties: nature, causes and treatment. *Current Perspectives in Psychological Science, 20*, 139–142.

Lyytinen, H., Erskine, J. M., Tolvanen, A., Torppa, M., Poikkeus, A.-M., & Lyytinen, P. (2006). Trajectories of reading development; a follow-up from birth to school age of children with and without risk for dyslexia. *Merrill-Palmer Quarterly, 52*, 514–46.

Glazzard, J. (2012). Dyslexia and self-esteem: Stories of Resilience. In T. Wydell (Ed.), *Dyslexia - A Comprehensive and International Approach*. Retrieved from: http://www.intechopen.com.

Mihandoost, Z., Elias, H., Sharifah, M. N., & Mahmud, R. (2011). A Comparison of the reading motivation and reading attitude of students with dyslexia and students without dyslexia in the elementary schools in Ilam, Iran. *International Journal of Psychological Studies, 3*, 17–27.

UNICEF (2003). *Examples of Inclusive Education.*, Retrieved from http://www.unicef.org/rosa/InclusiveSLK.pdf.

Ouafa Raziq

Université Hassan II, Morokko

Legasthenie in der marokkanischen Gesellschaft (Dyslexia in the Moroccan Society)

Abstract. Education and health are the basis of preparation for new perspectives in terms of overcoming difficulty of reading and writing for the future generations. With regard to the acquisition of minimum training, it is important for everyone to learn reading, writing and arithmetic in elementary school. It is necessary to detect disorders of the students in the aforementioned skills as early as possible to develop adaptive and necessary intervention programs for treatment. Based on this work, we would like to interview teachers in two different types of schools (public and private ones) in order to know whether teachers have ever known about "dyslexia" and what solutions they do if they recognize it in children and whether the Ministry of Education consider the diagnosis and the intervention to solve the problem.

Zusammenfassung. Bildung und Gesundheit sind die Basis jener Vorbereitung auf neue Perspektiven im Sinne der Überwindung von Lese- und Schreibschwierigkeit der nächsten Generation. Was den Erwerb der minimalen Ausbildung anbelangt, ist es wichtig für jeden Menschen das Lesen, Schreiben und Rechnen in der Grundschule zu lernen. Es ist notwendig die Früherkennung von Mängeln der Schülerinnen bei den o. g. Fertigkeiten festzustellen, um anpassende und erforderliche Interventionsprogramme für die Behandlung zu entwickeln. Wir führten Interviews mit Lehrern zweier verschiedener Schularten (staatliche und private Schule), um zu ermitteln, ob Lehrer überhaupt über Kenntnisse zur „Dyslexie" verfügen und welche Handlungsstrategien sie realisieren, wenn sie bei Kindern diese Diagnose erkennen und ob das Bildungsministerium Rücksicht auf das Problem der Diagnose und der Interventionsmöglichkeiten nimmt.

Keywords: Lese-Schreibschwierigkeit, Dyslexie, Lese-Rechtschreibfertigkeiten.

1 Einleitung

Bildung und Gesundheit sind die Basis jener Vorbereitung auf neue Perspektiven im Sinne der Überwindung von Lese-Rechtschreibschwierigkeit der nächsten Generation. Was den Erwerb der minimalen Ausbildung anbetrifft, ist es wichtig für jeden Menschen das Lesen und Schreiben in der Grundschule zu lernen. Da Lesen und Schreiben zu können ein Grundprinzip und eine Grundanforderung unserer Gesellschaft ist. Wer an Lese-Rechtschreibschwäche leidet, steht unter der Gefahr ins Abseits gedrängt zu werden. Füssenich (1999) geht davon aus, dass derjenige, der in der Schule lese-und rechtschreibschwach ist und bleibt, sich nach der Schulentlassung zu einem funktionalen Analphabeten entwickeln wird. Es ist notwendig, die Früherkennung von Mängeln der Schüler in den Lese- Rechtschreibfertigkeiten festzustellen, um anpassende und erforderliche Interventionsprogramme für die Behandlung zu entwickeln.

Man ästimiert, dass ungefähr 10 % der Weltbevölkerung von der Legasthenie betroffen sind. In Marokko gibt es leider keine offizielle spezifische Angabe über diese Pathologie. Der Versuch, genauere Informationen über Legasthenie in Marokko zu erhalten, wurde dadurch erschwert, dass selbst Pädagogen über fast keine Informationen verfügen. Man muss darauf verweisen, dass die ökonomischen, sozialen und kulturellen Faktoren als Ursachenerklärungen für die mangelnde Schriftsprachkompetenz in Marokko und Deutschland eine grundlegende Rolle spielen (Raziq, 2006). Auf diesem Grund werden wir anhand eines Interviews mit Lehrern in zwei verschiedenen Schularten, bzw. in zwei unterschiedlichen staatlichen und privaten Schulen zu wissen ob Schullehrer überhaupt Kenntnisse über „Dyslexie" verfügen und ob Kinder mit Lese-Rechtschreibschwierigkeiten gibt und welche Lösungsmöglichkeiten sie unternehmen, falls sie sie bei Kindern das Problem erkennen und ob das Bildungsministerium Rücksicht auf das Problem nimmt und Interventionsmöglichkeiten anbietet.

2 Die Relevanz der phonologischen und morphologischen Verarbeitung für den Schriftspracherwerb

Die phonologische Bewusstheit ist mit der Lesefertigkeit verbunden (Bradley & Bryant, 1983; Stanovich, Cunningham, & Cramer, 1984; Tunmer & Nesdale, 1985). Einige Wissenschaftler gehen von einer kausalen Beziehung zwischen phonologischer Bewusstheit und Lesen aus und betrachten die phonologische Bewusstheit als entscheidende Voraussetzung, als den ersten Schritt des Lesenslernens (Liberman & Liberman, 1990; Lundberg, Olofsson, & Wall, 1980; Mann & Liberman, 1984; Snowling, 1980). In den lateinischen Sprachen sind Vokale ein Teil des Alphabets, mit denen Wörter geschrieben werden. Der Fall ist jedoch in den semitischen Sprachen anders. So stellen die Kurzvokale keinen Teil des Alphabetes dar. Sie werden den Wörtern hinzugefügt, um deren Aussprache festzulegen und sind gewöhnlich nur für Anfänger gedacht. Texte werden ohne Kurzvokale für die normalen und kundigen Leser abgefasst (Abu-Rabia, 1997a, 1997b).

Für Lernanfänger des Arabischen werden die Kurzvokale geschrieben, weil sie eine entscheidende Rolle im Erlernen des Lesen und Rechtschreibens spielen (Abu-Rabia, 2002). Weiterhin haben sie eine zusätzliche Rolle hinsichtlich der

Identifizierung der grammatikalischen Funktion der Wörter in den Sätzen, die auf syntaktischen Kenntnissen basiert (Abu-Rabia, 2002). Eine Studie von Abu Rabia (1998) widmet sich der Untersuchung der Rolle der Phonologie, der verschiedenen Funktionen der Kurzvokale und der Rolle der Morphologie für die Lesegenauigkeit normaler und legasthenischer, arabisch sprechender Personen verschiedenen Alters beim Lesen arabischer Wörter und Sätze.

Charakteristisch für die arabische Sprache ist der Umstand, dass die Buchstaben miteinander innerhalb eines Wortes verbunden sind und ihre Form entsprechend ihrer Stellung im Wort sich ändert. Die meisten Wörter in der arabischen Sprache (Verben und Substantive) werden aus Wurzeln und Wortmustern konstruiert. Die Wurzeln bestehen aus 3 oder 4 Buchstaben. Die Wurzel der Morpheme stellt die grundlegende Einheit der Semantik des allgemeinen Wortes dar. Viele Wörter können von der Wurzel abgeleitet werden, z. B. die Wurzel / ktb (*Kataba*) = er schrieb; KTB (*kitab*) = Buch; KTB (*kutub*) = Bücher; KTB (*kitabahaa*) = ihr Buch; MKTB (*maktaba*) = Bibliothek; TKTB (*takataba*) = miteinander korrespondierte. Das Morphem-Muster liefert die exakte lexikalische Bedeutung und gleichzeitig die syntaktische Zuordnung. Zusätzlich bezieht sich das Morphem-Muster auf Person, Zahl, Geschlecht und Zeit.

Das größte Problem der legasthenen Kinder ist ein Defizit in der meta-phonologischen Kompetenz, welche zu großen Schwierigkeiten beim Lesen führt (Badda, Ahami, Gombert, El Qaj, Alami, & Lachheb, 2010). Legasthene Kinder leiden in der Tat unter einem Defizit in der Repräsentation im mentalen System und in der kognitiven Verarbeitung der Sprachlaute, die für das Erlernen von Graphem/Phonem-Korrespondenzen und ihre Handhabung während der Wiedergabe schädlich ist (Snowling, 2000; Sprenger-Charolles & Colé, 2003).

3 Das Schul-und Bildungssystem in Marokko

Will man über das marokkanischen Schul- und Bildungssystem heute sprechen, so muss man die gesellschaftliche Lage in Betracht ziehen. Angesicht der wirtschaftlichen Lage des Landes lebt ein großer Teil der Bevölkerung in Armut. So können viele Familien zwar ihre Kinder in die Schule schicken, jedoch müssen die Kinder die Schule frühzeitig verlassen, da die Schulen sowie die Kinder von dem Staat nicht genügend unterstützt werden. In den Schulen der armen Regionen findet man schlechte Bedingungen (Klassengröße, Schüler Lehrer-Relation, Schul- und Lehrausstattung) vor. Die Einschulungsquote beträgt offiziell landesweit ca. 80 %. Dennoch liegt die Prävalenz von Analphabetismus tatsächlich noch bei über 50 %. Sie ist auf dem Land deutlich höher als in den Städten. Besonders betroffen sind Frauen und Mädchen. Vor allem Mädchen auf dem Land haben trotz der 1963 eingeführten Schulpflicht keinerlei Schul-ausbildungen.

Im Vergleich zu Deutschland lag Marokko unterhalb des internationalen Mittelwertes. Während Deutschland im Mittel bei 539 und damit auf Platz 11 lag, konnten in Marokko nur 350 Punkte und damit der vorletzte Platz erreicht werden. Als Schlussfolgerung kann man davon ausgehen, dass ungünstige soziale Bedingungen, mangelnde Verbreitung von Büchern, geringe Motivation der Schüler und eine zunehmende Kluft des Marokkanischen Dialektes zur arabischen Sprache eine große Rolle spielen. Arabisch ist für viele Kinder keine Muttersprache. In der Interpretation wird betont, dass der Erwerb formaler Sprache weniger Funktionalität im Alltagsgebrauch hat (Mullis, Martin, Gonzalez, & Kennedy, 2003).

4 Zum Begriff Legasthenie: Historische Aspekte

Charakteristisch für die Lese-Rechtschreibschwäche (LRS) ist der Umstand, dass die Betroffenen Leserechtschreibfähigkeiten verzögert und erschwert erwerben. Diese Probleme können nicht auf alternative Ursachen zurückgeführt werden, wie mangelhafte Beschulung, emotionale bzw. neurologische Störungen und Intelligenzminderung. Das Problem des Schriftspracherwerbes wurde zum ersten Mal am Ende des 19. Jahrhundert erörtert. Der Begriff Legasthenie wurde von dem ungarischen Neurologen Paul Ranschburg (1916) gebraucht (in Warnke, 1990). Ranschburg sprach zum ersten Mal vom Aspekt phonologischer Defizite, indem er nicht nur auf visuelle Verwechselungsfehler, sondern auch auf solche, phonologischer Art bezogen auf ähnliche Laute hinwies. Maria Linder (1951) definierte Legasthenie, indem sie sich auf die Gedanken von Morgans stützte, als Teilleistungsstörung. Seitdem findet der Begriff Legasthenie seinen Platz in der wissenschaftlichen Forschung. Es gibt unterschiedliche Definitionen zum Terminus Legasthenie. So reichen die verwendeten Begriffe von Störung der Schriftsprache (u. a. Berkhan, 1885; Ranschburg, 1916) bis zu entwicklungsbedingter Lesestörung (Koeler & Sass, 1904, in Warnke, 1990).

In anderen Ländern spricht man von Dyslexia oder Dyslexie. Auch im arabischen Sprachraum gibt es Definitionen, die sich nicht von denen aus England und Frankreich unterscheiden. In einer Studie schreibt Alhamdan (1992, in Taha, 2002), dass die Leseschwäche in der arabischen Sprache durch folgende Kriterien charakterisiert ist: (1) Mangel an Grundlagen des Lesens, was auch die Fächer betrifft, die nicht direkt das Lesen zum Gegenstand haben wie Geometrie und Mathematik; (2) Die Unfähigkeit, zwischen den verschiedenen Buchstaben zu unterscheiden, (3) Vernachlässigung und Weglassen wichtiger grundlegender Bestandteile des Schreibens wie Punkte und Verwechslung ähnlicher Buchstaben-

formen (س=S) (س=s: stimmloser dentaler Reiblaut) und (ش=sch) (ش=š: stimmm-
loser palato - alveolar).

5 Methoden

Im Rahmen dieses wissenschaftlichen Projektes haben wir versucht, einen In-
terviewleitfaden zu entwickeln. Die Interviews, welche wir mit Lehrern aus verschiedenen Institutionen durchgeführt haben, wurden danach ausgewertet, um
zu sehen, ob und wie gut sich Lehrer mit dem Legasthenie-Phänomen auskennen.

6 Ergebnisse

Die Mehrheit der befragten Lehrer ist mit dem Legasthenie-Phänomen nicht
vertraut. Die Lehrer, die in den privaten Elite-Schulen in Casablanca arbeitet und
welche mit Psychologen zusammenarbeitet, wissen, was Dyslexia bedeutet. Sie
behaupten, dass ihre Institution über keine Instrumente verfügt, welche sie beim
Umgang mit den betroffenen Kindern unterstützen könnten. Sie bestätigen, dass
das Schulbildungsministerium nicht in irgendeiner Weise interveniert bezüglich
die Schwierigkeiten bei Schülern im Lesen und Schreiben. Sie behaupten auch,
dass das Ministerium, keine Rücksicht auf diese Problematik nimmt und bietet
auch keine pädagogischen und didaktischen Methoden für diese Schülern an,
um das Problem zu mindern. Das Ministerium bietet keine spezielle Ausbildung
für Lehrer, um eine wirksame Hilfe für diese Kinder zu erreichen. Sie behaupten,
dass durch das Ministerium keine spezielle Förderung bei Schreib- und Lese-
schwierigkeiten als auch keine speziellen Klassen für solche Kinder zur Verfügung
gestellt werden. Die Lehrer hoffen, dass das Schulbildungsministerium päda-
gogische Programme im Rahmen der gesamten akademischen Schullaufbahn
integriert. Da die Problematik der Lese-Rechtschreibschwäche kaum Beachtung
in der arabischen Welt erfährt und demnach keine wissenschaftlich fundierten
psychodiagnostischen Testverfahren existieren. So sind Legastheniker trotz einer
guten Intelligenz und eines normalen Hörvermögens konfrontiert mit Misserfolg
wegen ihrer extremen Schwierigkeiten, geschriebene Wörter zu erkennen. Es ist
wichtig zu betonen, dass die Verbesserung der Schulen sich als eine Notwendig-
keit für die Entwicklung und Gesundheit des Kindes erweist (Ahami et al., 2006;
Badda, 2008).

7 Diskussion

Zusammenfassend lässt sich anhand unserer Interviews konstatieren, dass
es kaum Beachtung und Unterstützung von der Seite der Pädagogen und des

Schulbildungsministeriums gibt. In Marokko wird durch das Schulbildungssystem die Existenz dieser kognitiven Störung nicht anerkannt. Aber einige Studien aus Marokko weisen indirekt auf das Phänomen „Legasthenie" bei Kindern mit arabischer Muttersprache als auch bei Kindern mit französischer Sprache als zweiter Fremdsprache hin (Badda, 2008).

Lesen ist eine komplexe kognitive Aktivität erster Ordnung (Gombert, 1990). Es ist eine geistige Fähigkeit, die sich durch mehrere Komponenten unterschiedlich, aber ergänzend ergibt. Das Lesenlernen ist die Grundbasis der ersten Schuljahre. Darüber hinaus bleibt das Lesen eines der vorrangigen Ziele der Grundschule. Die Bedeutung dieser Tätigkeit für die Schule, für das soziale Leben und für die berufliche Tätigkeit ist weit verbreitet. In der Tat kann man sagen, dass eine Person in jeder Gesellschaft und in jeder Schule funktionsfähig sein kann, erst wenn sie lesen kann. Man muss darauf hinweisen, dass die Schwierigkeiten beim Lesenlernen erhebliche Auswirkungen auf andere Schulfächer haben können (Demont & Gombert, 2004). Es gibt keine klare und logische Interpretation auf der Seite der meisten Lehrer betreffend dieses Problems der Lese und Rechtschreibschwierigkeiten. Selbst die Pädagogen verfügen über keine ausreichenden Kenntnisse hinsichtlich dieser Problematik.

In unserer marokkanischen Gesellschaft als auch in der Schule wird Legasthenie noch immer mit Dummheit, Unkonzentriertheit und Faulheit verwechselt. Daher kann man sagen, dass das Wissen um das Phänomen leider fehlt und man damit nicht bewusst umgehen kann. Hinzu kommt, dass es keine spezialisierten Fachleute für Legasthenie gibt. Deswegen gehen wohlhabende Eltern betroffener Kinder zum Ergotherapeuten oder zum Logopäden entweder überwiesen vom Psychiater oder vom Arzt.

Unser Ziel sollte demzufolge darin bestehen, Kinder mit Schwierigkeiten beim Lesen- und Schreibenlernen frühzeitig zu erkennen, um durch gezielte Präventionsmaßnahmen Spätfolgen, wie im schlimmsten Fall Analphabetismus zu vermeiden und den Lehrern bewusst über die Anwesenheit diese Problematik machen.

8 Kontakt

Dr. Ouafa Raziq
Institution: Université Hassan II, Ain Chock
Address: 17 Rue Amyot, Quartier des Hôpitaux, App 5 Casablanca, Marokko
E-mail: ouafaeadam@hotmail.fr

9 Literatur

Abu-Rabia, S. (1997a). Reading in Arabic orthography: The effect of vowels and context on reading accuracy of poor and skilled native Arabic readers. *Reading and Writing: An Interdisciplinary Journal, 9*, 65–78.

Abu-Rabia, S. (1997b). Reading in Arabic orthography: The effect of vowels and context on reading accuracy of poor and skilled native Arabic readers in reading paragraphs, sentences and isolated words. *Journal of Psycholinguistic Research, 26*, 465–482.

Abu-Rabia, S. (1998). Reading Arabic texts: Effects of text type, reader type, and vowelization. *Reading and Writing: An Interdisciplinary Journal, 10*, 106–119.

Abu-Rabia, S. (2002). Reading in a root-based morphology language. *Journal of Research in Reading, 25*, 299–309.

Ahami, A. O. T., Mari, E., Aboussalah, Y., Biyadi, N., Azzaoui, F. Z., Samih, M., Mesfioui, A., & Elbouhali, B. (2006). *Dépistage des troubles psychomoteurs et comportementaux chez le jeune enfant.* Santé, Education et Environnement, 33–41. Digi Edition, ISBN: Rabat, Maroc.

Badda, B. (2008). Apprentissage de la lecture, dyslexiephonologiqueetremédiation par lelogiciel « ItinéraireCombinatoire » chez l'enfantmarocain. *Thèse de doctoratCotutelle*, Université Ibn TofailMaroc – Université de Rennes 2 France.

Badda, B., Ahami, A. O. T., Aboussaleh, Y., Bahtit, J., & Gombert, J.-E. (2008). Repérage des difficultés d'apprentissage de lecture évoquant la dyslexie phonologique chez des enfants marocains. *Pathologies Humaines et Déficits de Développement Approche Pluridisciplinaire.* Edition Imprimerie Rapide : 33–37.

Badda, B., Ahami, A. O. T., Gombert, J.-E., El Qaj, M., Alami, N., & Lachheb, A. (2010). Enfants marocains scolarisés: Essai de remédiation de la dyslexie phonologique via le logiciel « Itinéraire Combinatoire », *EpiNet: la revue électronique de l'EPI*, n° 129. Retrieved from http://www.epi.asso.fr/revue/articles/a1011c.htm.

Berkhan, O. (1885). Über die Schriftsprache bei Halbidioten und ihre Ähnlichkeit mit dem Stammeln. *Archiv für Psychiatrie, 16*, 78–86.

Berkhan, O. (1886). Über die Schriftsprache bei Halbidioten und ihre Ähnlichkeit mit dem Sprachgebrechen. 2. Stammeln und Stottern. *Archiv für Psychiatrie, 17*, 861–897.

Bos, W., Lankes, E. M., Prenzel, M., Schwippert, K., Walter, G., & Valtin, R. (Hrsg.) (2003). Erste Ergebnisse aus IGLU. *Schülerleistungen am Ende der vierten Jahrgangsstufe im internationalen Vergleich.* Münster: Waxmann.

Bradley, L., & Bryant, P. (1983). Categorizing sounds and learning to read: A causal connection. *Nature, 301*, 419–421.

Braibant, J. E. (1994). Le décodage et la compréhension. *In* Grégoire, J., Piérart, B. (1994). *Évaluer les troubles de la lecture.* Bruxelles : De Boeck.

Casalis, S. (1995). *Lecture et dyslexies de l'enfant.* Paris : Presses universitaires du Septentrion.

Demont, E., & Gombert, J. E. (2004). L'apprentissage de la lecture : évolution des procédureset apprentissage implicite. *Enfance, 3,* 245–257.

Füssenich, I. (1999). Funktionaler Analphabetismus in Deutschland. In J. E. Gombert, P. Colé, S. Valdois, R. Goigoux, P. Mousty, & M. Fayol (Eds.), *Enseignerla lecture au cycle 2,* (pp. 183–188). Paris: Nathan Pédagogie.

Kail, M., & Fayol, M. (2000). *L'acquisition du langage : le langage en développement audelàde trois ans.* Paris: Presses universitaires de France.

Liberman, I. Y., & Liberman, A. M. (1990). Whole word vs. code emphasis: Underlying assumptions and their implication for reading instruction. *Bulletin of the Orthon Society, 40,* 51–76.

Lundberg, I., Olofsson, A., & Wall, S. (1980). Reading and spelling skills in the first school years predicted from phonemic awareness skills in kindergarten. *Scandinavian Journal of Psychology, 21,* 159–173.

Mann, V. A., & Liberman, I. Y. (1984). Phonological awareness and verbal short-term memory: Can they presage early reading problems? *Journal of Learning Disabilities, 17,* 592–599.

Raziq O. (2006). *Verteilung der sozioökonomischen Status von Legasthenikern und Normallesenden.* Unpubliziertes Manuskript.

Snowling, M. (1980). The development of grapheme-phoneme correspondence in normal and dyslexic readers. *Journal of Experimental Child Psychology, 29,* 294–305.

Snowling, M. (2000). *Dyslexia* (2nd ed.). Oxford: Blackwell.

Sprenger-Charolles, L., & Colé, P. (2003). *Apprentissage de la lecture et Dyslexie: Desrecherches fondamentales aux implications pratiques.* Paris: Dunod.

Stanovich, K. E., Cunningham, A. E., & Cramer, B. B. (1984). Assessing phonological awareness in kindergarten children: Issues of task comparability. *Journal of Experimental Child Psychology, 38,* 175–190.

Taha, D. (2002). *Schwierigkeiten beim Lesen und Schreiben der arabischen Sprache bei bilingualen Schulern.* (Unpublished master thesis). Alchams University, Cairo, Egypt.

Tunmer, W. E., & Nesdale, A. R. (1985). Phonetic segmentation skill and beginning reading. *Journal of Educational Psychology, 77,* 417–427.

Warnke, A. (1990). *Legasthenie und Hirnfunktion: Neuropsychologische Befunde zur visuellen Informationsverarbeitung.* Bern: Verlag Hans Huber.

Adil Ishag

International University of Africa, Sudan

Diglossische Aspekte beim Arabischlernen (Diglossic Aspects in Arabic Language Learning)

Abstract: Diglossia emerges in a situation where two distinctive varieties of a language are used alongside within a certain community. In this case, one is considered as a high or standard variety and the second one as a low or colloquial variety. Arabic is an extreme example of a highly diglossic language. This diglossity is due to the fact that Arabic is one of the most spoken languages and spread over 22 Countries in two continents as a mother tongue, and it is also widely spoken in many other Islamic countries as a second language or simply the language of Quran. The geographical variation between the countries (where the language is spoken and the duality of the classical Arabic and daily spoken dialects in the Arab world), which makes the Arabic language to one of the most diglossic languages in the world. This paper tries to investigate this phenomena and its relation to learning Arabic as a first and second language.

Zusammenfassung. Diglossie entsteht in einer Situation, in der zwei verschiedene Varianten einer Sprache nebeneinander innerhalb einer bestimmten Gemeinschaft benutzt werden. In diesem Fall wird eine als Hoch- oder Standard-Variante und die zweite als niedrige oder umgangssprachliche Variante benutzt. Arabicum ist ein extremes Beispiel einer hoch diglossischen Sprache. Diese Diglossie entsteht aufgrund der Tatsache, dass Arabisch eine der am meisten gesprochenen Sprachen ist und in mehr als 22 Ländern, auf zwei Kontinenten als Muttersprache verbreitet ist. Darüber hinaus es wird auch in vielen islamischen Ländern als Zweitsprache oder einfach als die Sprache des Korans benutzt. Die geografischen Unterschiede zwischen den Ländern, in denen die Sprache gesprochen wird, und die Dualität zwischen dem klassischen Arabisch und den täglich gesprochenen Dialekten in der arabischen Welt macht die arabische Sprache zu einer der stärksten diglossischen Sprachen der Welt. Dieser Artikel versucht, diese Phänomene und ihre Beziehung zum Arabisch-Lernen als erste und zweite Sprache zu untersuchen.

Keywords: Diglossie, Arabischlernen, H-Varietät, L-Varietät, Spracherwerb.

1 Einführung in die Thematik

Der soziolinguistische Terminus Diglossie wurde von Ferguson (1959) wieder eingeführt und aufgegriffen. Diglossie entsteht in einer Situation, wobei zwei unterschiedliche Sprachformen innerhalb einer bestimmten Sprachgemeinschaft verwendet werden. In diesem Fall wird eine Varietät als Hochvarietät bezeichnet (engl. High variety) und die andere Varietät wird als niedrige Varietät (engl. Low variety) bezeichnet. Arabisch ist ein extremes Beispiel für eine hochdiglossische Sprache und von daher wurde Arabisch von Ferguson neben Griechisch, Schweizerdeutsch, haitianisches Kreol, als anschauliche Beispiele für Diglossie vorgestellt. Arabisch wird allerdings als extremdiglossiche Sprache betrachtet (al-Batal, 1992). Im Arabischen wird Hocharabisch (Fusha) als die Hochvarietät (H-Varietät) und Standardsprache betrachtet, wohingegen das umgangssprachliche und regional geprägte Arabisch (Ammiya) als niedrige Varietät (L-Varietät) bezeichnet wird.

Hierbei kann man zwischen Diglossie und Bilingualismus unterscheiden. Diglossie bezieht sich auf die gesamte Sprachgemeinschaft, während Bilingualismus sich auf die individuelle Ebene bezieht (Mackey, 1989). Die geografischen Unterschiede zwischen den Ländern, in denen die Sprache gesprochen wird und die Dualität des klassischen Arabisch und der alltäglich gesprochenen Dialekte in der arabischen Welt auf der anderen Seite, führen zu Komplexeren Schwierigkeiten beim Arabischlernen. Dieser Aufsatz versucht dieses Phänomen im Zusammenhang zum Arabischlernen als Erste- und als Fremdsprache zu diskutieren.

2 Merkmale der Diglossie

Bei einer Diglossischen Situation kann man folgende Charakteristika erkennen: Diglossie ist eine relativ stabile Sprachsituation. Beispielsweise reicht die Diglossie im Arabischen und der Konflikt zwischen der Standardsprache und den Dialekten ursprünglich bis Jahrhunderte zurück in die Vergangenheit, und besteht immer noch bis in die Gegenwart.

Die Hochvarietät ist grammatisch, syntaktisch, lexikalisch und stilistisch wesentlich komplexer als die niedrige Varietät. In diesem Fall weist Hocharabisch mehr grammatische Kategorien als die Dialekte auf, sowie Casus, Numerus und Genus. Außerdem ist die Hochvarietät kodifiziert und normiert. Die H-varietät verfügt über zahlreiche Lehrmaterialen und Grammatikbücher. Für den Dialekt wird aber kein metasprachliches Wissen entwickelt.

Die Hochvarietät genießt höheres Ansehen als die niedrige Varietät. Hierbei wird Hocharabisch als prestigeträchtige Sprache betrachtet, während die arabischen Dialekte als abweichend betrachtet werden. Darüber hinaus argumentiert Palmer (2007), dass „gesprochenes Arabisch oft stigmatisiert und als weniger prestigeträchtig betrachtet wird und von daher auch wenige Untersuchungen über das gesprochene Arabisch durchgeführt werden, obschon sie als Umgangssprache für die Alltagskommunikation gilt.

Jeder Sprachvarietät wird eine bestimmte Funktionalität zugewiesen. Nachrichten und die gedruckten Medien, sowie Zeitungen und Zeitschriften erscheinen fast ausschließlich auf Hocharabisch. In umgangssprachlichen Situationen wird dagegen der jeweilige Landesdialekt verwendet.

Die Standardvarietät trägt das literarische Erbe der Sprachgesellschaft. Hierbei wird die geschriebene Literatur, wie z. B. Sachbücher, sämtlich auf Hocharabisch verfasst. Der Dialekt hat grundsätzlich keine selbstständige schriftliche Form.

Die Standardsprache wird erst in der Schule gelernt, wohingegen der Dialekt früh als Muttersprache automatisch erworben wird. Bezüglich Arabisch wird Hocharabisch erst in der Schule durch den Unterricht vermittelt und es ist niemandes Muttersprache. Von daher wird Hocharabisch nur von gebildeten Leuten gesprochen. Auf der anderen Seite wird der regional geprägte Dialekt durch Sozialisation erworben und gilt als Muttersprache derer Sprecher.

3 Die arabische Sprache und deren dialektale Aufteilung

Arabisch ist eine der 6 offiziellen Sprachen der Vereinten Nationen und die fünftgrößte Sprache der Welt. Sie gehört zu der semitischen Sprachfamilie, zu der auch Hebräisch und Amharisch gezählt werden und ist die meistverbreiteste semitische Sprache. Arabisch wird heute von ca. 422 Millionen Menschen als Mutter- und Zweitsprache in 22 arabischen Ländern und anderen Ländern in Asien und Afrika gesprochen und gilt als Sprache der koranischen Offenbarung. Im Arabischen kann man 3 unterschiedliche Sprachvarianten unterscheiden nämlich; Hocharabisch (Klassisches Arabisch wie im Koran: CA), Das moderne Standardarabisch (MSA) und Dialektalarabisch.

Das moderne Hocharabisch gilt als Standardsprache und dominiert den schriftlichen Bereich. Sie ist in allen arabischen Ländern gleich. Dialektalarabisch ist geographischbedingt und variiert je nach Land und Gebiet. Der Unterschied zwischen Hocharabisch und den einzelnen arabischen Dialekten ist phonetisch, syntaktisch, morphologisch und lexikalisch zum Teil erheblich.

Die arabischen Dialekte können grundsätzlich in 6 Gruppen untergeteilt werden:

a. Halbinsel Arabisch: umfasst Golf Arabisch und Jemenitisches Arabisch.

b. Levantinisches Arabisch: umfasst Syrien, Jordanien, Libanon, und palästinensisches Arabisch.

c. Mesopotamisches oder irakisches Arabisch.

d. Maghribisches Arabisch: umfasst Marokko, Tunesien, Algerien, Libyen, und Mauretanien.

e. Ägyptisches Arabisch: umfasst auch zum Teil Nordsudan zumindest auf die semantische Ebene.

f. Sudanesisches Arabisch: umfasst zum Teil Tschad, Eritrea und Somalia.

Khaled (2010) geht davon aus, man dass im Bezug auf die phonetischen, stilistischen, und linguistischen Aspekte in einem groben Trennungsversuch den gesamten arabischen Sprachraum in 2 großen Gruppen aufteilen kann: die westarabischen Dialekte: die in Marokko, Tunesien, Algerien, Libyen, Mauretanien gesprochen werden und die ostarabischen Dialekte: die die anderen arabischen Dialekte umfasst.

Aufgrund dessen, dass Ägypten das bevölkerungsreichste Land in der arabischen Welt ist und den Medien- und Filmbereich dominiert, ist ägyptisch der meistverwendete Dialekt und wird heutzutage überall verstanden. Jedoch wird levantinisches Arabisch neben ägyptischem Arabisch am meisten von den Arabischstudierenden als Fremdsprache bevorzugt. Außerdem werden die Dialekte zum Teil von den ehemaligen Kolonialsprachen wie Englisch oder Französisch beeinflusst. Darüber hinaus, wird aufgrund der breiten sprachlichen Distanz zwischen H-Varietät und L-Varietät, insbesondere auf der syntaktischen und morphologischen Ebene, die Frage gestellt, ob die dialektalen Mundarten überhaupt als Arabisch betrachtet werden können oder sie einfach Manifestationen der lokalen nationalen Kultur sind (Haeri, 2000).

4 Das Schriftsystem und Arabischlernen

Die arabische Schrift ist hauptsächlich eine Konsonanzschrift und ist die zweitverbreitete Schrift nach Lateinisch. Die ähnlichen Buchstaben werden durch diakritische Punkte unterschieden und differenziert. Ein wesentlicher Nachteil der arabischen Schrift besteht im Fehlen der Buchstaben für Kurzvokale. Das Fehlen von Vokalzeichnen (Harakat) im arabischen ist idiosynkratisch zum arabischen Schriftsystem. Deshalb werden Kurzvokale mittels Zusatzzeichen (Harakat) repräsentiert. Außerdem sind die Buchstaben von ihrer Position im Wort (Anfang, Mitte, Ende, oder isoliert sowie die Anzahl der Punkte) abhängig. Das Problem besteht darin, dass bei gedruckten Medien und Bücher diese Vokalzeichnen total weg gelassen werden. Dies erschwert den Leseprozess und die Sprachverarbeitung im Arabischen und gilt als eines der größten Hindernisse beim Arabischlernen. Ein Wort wie (Din) zum Beispiel kann im arabischen unterschiedlich interpretieren werden je nach Kontext. Es kann Religion oder Schuld bedeuten. Wenn solche Wörter als Homographie isoliert ohne Vokalzeichnen stehen, wird es dann schwer die Bedeutung zu erraten.

5 Lesekompetenz beim Arabischerwerb

Im Arabischen muss der Leser verstehen um lesen zu können, während in fast allen anderen Sprachen die Menschen lesen um zu verstehen. Deswegen scheint der Leseprozess im Arabischen eine umgekehrte Aufgabe zu sein. Der linguistische Abstand zwischen geschriebenem und gesprochenem Arabisch wirkt negativ auf die Entwicklung der Lesekompetenz (Hansen, 2014; Saiegh-Haddad, 2003). Darüber hinaus haben arabische Kinder Schwierigkeiten beim Leseerwerb. Dies wurde durch ihre relativ niedrige Leistung in den internationalen Lesetests reflektiert (Pirls, 2006). Asadi und Ibrahim (2014) gehen davon aus, dass dieses niedrige Niveau der Lesekompetenz unter der arabischsprachigen Bevölkerung mit der Komplexität der Orthographie und dem dichten morphologischen und syntaktischen System im Arabischen und deren Diglossiesituation zurückzuführen ist.

Arabische Kinder lesen langsamer und machen mehr Fehler im Vergleich zu Kindern, die in einigen anderen Sprachen lesen (Saiegh-Haddad, Levin, Hende, & Ziv, 2011). Weiterhin, zeigten einige Studien (Eviatar & Ibrahim, 2014) dass arabische Studenten mehr Fehler machen und im Vergleich zu Studenten in anderen Sprachen wie z. B. in Romanischen Sprachen langsamer lesen. Auch die erfahrenen Erwachsenen sind zum Teil davon betroffen.

Lesen im Arabischen ist ein mehrstufiger Prozess; der Leser muss erst den Text entziffern; die Kurzvokale voraussagen und dabei mehrere alternative Wörter im Arbeitsgedächtnis verarbeiten und testen, um sich für eine Bedeutung entscheiden zu können. Dies impliziert, dass der arabische Leser die Wörter schneller identifizieren muss, als in anderen Skripten um den Sinn eines Texts zu finden. Das scheint aber nicht in Wirklichkeit der Fall zu sein, denn sie identifizieren die Wörter langsamer und machen mehr Fehler.

6 Diglossie beim Arabischlernen

Diglossie hat nicht nur für Ausländer Nachteile beim Arabischlernen als Fremdsprache, sondern auch für die vermeintlichen arabischen Muttersprachler als Erstsprache. Die Auffassung, dass Arabisch für die Araber eine Fremdsprache ist oder eine Art Quasi-Fremdsprachenstatus in der arabischen Welt hat, wird schon lange diskutiert. Hierzu schrieb Maamouri (1998), dass das moderne Standardarabisch Niemandes Muttersprache ist und nur selten oder fast nie zu Hause in der arabischen Welt gesprochen wird. Der Dialekt wird schon in der Kindheit erworben, während Hocharabisch erst in der Schule gelernt wird. Arabisch wird von Kindern an der Schule wie eine Fremdsprache verarbeitet und sie verhalten sich dazu sprachlich und metasprachlich wie Bilinguale (Eviatar & Ibrahim, 2014).

In einer psycholinguistischen Studie von Ibrahim und Aharon-Peretz (2005) wurde gezeigt, dass trotz der gemeinsamen Herkunft des gesprochenen und geschriebenen Arabisch und des täglichen intensiven Gebrauchs beider Varietäten von den arabischen Einheimischen, beide Sprachen den Status als erste und zweite Sprache im kognitiven System beibehalten.

In einer weiteren ähnlichen Studie von Ibrahim (2009) hat der Forscher ein Experiment über lexikalische Entscheidung zwischen palästinensisch gesprochenem Arabisch, Standardarabisch und Hebräisch durchgeführt. Die Ergebnisse zeigten, dass der Status von Standardarabisch zu dem Status von Hebräisch ähnlich ist und der typischen Organisation einer Zweitsprache in einem separaten Lexikon ähnelt. Somit scheint Arabischlernen in mancher Hinsicht eher wie das Erlernen einer zweiten Sprache, als das Erlernen der formellen Register der eigenen Muttersprache zu sein. Weiterhin zeigten Fedda und Oweini (2012), dass Diglossie die Wortschatzentwicklung bei jungen arabischen libanesischen Studenten behindert. Zahlreiche Studien belegten, dass Schüler Schwierigkeiten beim Erlernen der arabischen Sprache haben, die in erster Linie auf die diglossische Natur im Arabischen zurückzuführen sind (Ayari, 1996). Außerdem haben arabische Kinder vor allem beim gesprochenen Hocharabisch Erwerbsprobleme, was auf fehlendes gesprochenes Hocharabisch im Alltag zurückzuführen ist.

In diesem Zusammenhang versucht dieser Aufsatz, mögliche Schlussfolgerungen beim Arabischlernen abzuleiten:

Da Standardarabisch hauptsächlich an der Schule gelernt wird, wird es tendenziell nur von gebildeten Leuten gesprochen. Die Araber können zwar Hocharabisch gut verstehen, sie können sich aber nicht in Hocharabisch fließend verständigen oder schriftlich ausdrücken. Sie haben eher ein passives Wissen bzw. passive Fertigkeiten (Lesen und Hören), das Problem ist eher bei der aktiven Sprachproduktion (Sprechen und Schreiben).

Hocharabisch wird nur in der Schule oder durch Unterricht beigebracht. Deswegen haben mehr als 120 Millionen Araber nicht die Möglichkeit, die H-Form zu lernen, was insgesamt zu einer Alphabetisierungsrate in Nordafrika und im Nahen Osten von ca. 60 % führt (Bani-Khaled, 2014).

Eine muttersprachliche Beherrschung des Arabischen scheint unplausible zu sein, wenn es heutzutage keinen Arabischmuttersprachler gibt. Auch die gebildeten Araber sprechen kaum vollständig Hocharabisch. Sie neigen eher dazu, eine mittlere Version der Sprache zwischen Standard- und Dialektarabisch zu sprechen oder code-switching (Sprachwechsel) zwischen den beiden Varietäten zu praktizieren. Von daher werden manchmal grammatische Unkorrektheiten und sprachliche Defizite im Arabischen toleriert sowohl von den vermeintlichen Muttersprachlern als auch von den Fremdsprachlern.

Hocharabisch unterscheidet sich von Umgangsarabisch in vier Hauptteilen der Sprache: Wortschatz, Phonologie, Syntax und Grammatik (Abu-Rabia, 2000). Deshalb ist das Erlernen von den beiden Varietäten H und L beim Arabischlernen erforderlich. Eine Varietät kann die andere überhaupt nicht ersetzen, um in der arabischen Welt zurecht zu kommen, und die daraus entstehenden Kommunikationsprobleme im Alltag zu vermeiden.

Ganz im Allgemeinen könnte gesagt werden, dass Verständigungsschwierigkeiten binnenarabischen auftreten, wenn ein geographischer Abstand zwischen den dialektalen Varietäten besteht. Ein Syrer verständigt sich z. B. müheloser mit einem Jordanier als mit einem Tunesier. Jedoch bestehen zwischen den unterschiedlichen Dialekten, was zum Teil die abstrakten Themen wie Politik und Religion betrifft, erhebliche Gemeinsamkeiten.

7 Kontakt

M.A. Adil Ishag
Institution: International University of Africa, Sudan
Address: Madani ST. Khartoum 12223, Sudan
E-mail: adilishag@iua.edu.sd

8 Literatur

Abu-Rabia, S. (2000). Effects of exposure to literary Arabic on reading comprehension in a diglossic situation. *Reading and Writing: Interdisciplinary Journal, 13*, 147–157.

Al-Batal, M. (1992). Diglossia proficiency: the need for an alternative approach to teaching. In A. Rouchdy (Ed.), *The Arabic Language in America* (pp. 284–304). Detroit, Michigan: Wayne State Press.

Asadi, I., & Ibrahim, R. (2014). The influence of diglossia on different types of phonological abilities in Arabic. *Journal of Education and Learning, 3*(3), 45–55.

Ayari, S. (1996). Diglossia and illiteracy in the Arab World. *Language, Culture and Curriculum, 9*(3), 243–252.

Bani-Khaled, T., A. (2014). Standard Arabic and diglossia: A problem for language education in the Arab World. *American International Journal of Contemporary Research, 4*(8), 180–189.

Eviatar, Z., & Ibrahim, R. (2014). Why is it hard to read Arabic? In Saiegh-Haddad, E. & Joshi, M. (Eds.), *Handbook of Arabic Literacy: Insights and Perspectives. Literacy Studies 9* (pp. 77–98). New York, NY: Springer.

Fedda D. O., & Oweini, A. (2012). The effect of diglossia on vocabulary acquisition in Arabic of Lebanese students. *Educational Research and Reviews, 7*(16), 351–361.

Ferguson, C. A. (1996). Diglossia. In T. Huebner (Ed.), *Sociolinguistic Perspectives: Papers on Language in Society 1959–1994* (pp. 25–39). New York: Oxford University Press.

Haeri, N. (2000). Form and ideology: Arabic sociolinguistics and beyond. *Annual Review of Anthropology, 29*, 61–87.

Hansen, G. F. (2014). Word recognition in Arabic: approaching a Language-specific model. In Saiegh-Haddad, E. & Joshi, M. (Eds.), *Handbook of Arabic Literacy: Insights and Perspectives. Literacy Studies 9*, (pp. 55–76). New York, NY: Springer.

Ibrahim, R., & Aharon-Peretz, J. (2005). Is literary Arabic a second language for native Arab speakers? Evidence from a semantic priming study. *The Journal of Psycholinguistic Research, 34*(1), 51–70.

Ibrahim, R. (2009*)*. The cognitive basis of diglossia in Arabic: Evidence from a repetition priming study within and between languages. *Psychology Research and Behavior Management, 2*, 93–105.

Maamouri, M. (1998). Language education and human development: Arabic diglossia and its impact on the quality of education in the Arab region. *Mediterranean Development Forum*. Morocco: World Bank.

Mackey, W., F. (1989). Le genèse d'une typologie de la diglossie. *Revue Québécoise de Linguistique Théorique et Appliquée, 8*(2), 11–20.

Palmer, J. (2007). Arabic diglossia: Teaching only the standard variety is a disservice to students. *Arizona Working Papers in SLA & Teaching, 14*, 111–122.

Saiegh-Haddad, E. (2003). Linguistic distance and initial reading acquisition: the case of arabic diglossia. *Applied Psycholinguistics, 24*(3), 431–451.

Saiegh-Haddad, E., Levin, I., Hende, N., & Ziv, M. (2011). The linguistic affiliation constraint and phoneme recognition in diglossic Arabic. *Journal of Child Language, 38*(2), 297–315.

Guangshu Gu[1] & Dian Sari Utami[2]

[1] Wuhan Textile University, China

[2] Islamic University of Indonesia, Indonesia

Family Elements of Reading Problems among Children in a Chinese Environment

Abstract. Chinese has a morph syllabic writing system, which makes it different and unique compared to Western languages. The character and basic units represent a morpheme as well as a syllable. Chinese children learn this language in the family as the first and basic element to learn and imitate. This article will further explain the family elements as both risk and protective factors that are assumed to influence the reading problems among children in a Chinese environment.

Keywords: family elements, Chinese language, children's reading problems.

1 Introduction

Until several decades ago, it was believed that children's reading problems existed only in Western languages. Chinese, which is different from the Western alphabetic writing system, is a morph syllabic writing system. Its basic unit and character represent a morpheme as well as a syllable. Although Chinese is the native language of the largest population in the world, there is relatively little research on Chinese reading problems until the end of the 1970s (Shu, Meng, Chen, Luan, & Cao, 2005).

Family is the first place for children to learn, and parents are the first teachers for the children to learn and imitate. In children's learning, a significant effect occurs from the family's reading habits, the reading interactions among family members, their socio-economic status, the collection of books for children, and the reading environment. Several studies have demonstrated that Chinese dyslexic children have lower phonological awareness compared to normally achieving children.

Dyslexia is among the most common neurodevelopmental disorders, with a prevalence of 5 %–12 % (Katusic, Colligan, Barbaresi, Schaid, & Jacobsen, 2001). There are other reading difficulties that are unrelated to dyslexia. Chinese literature reports the prevalence rate is 3.26 %. At least 50 % of children with learning disabilities are children with reading problems (Chen, 2007). Depending on the phenotype dimension investigated, inherited factors are estimated to account for up to 80 % of these problems (Schumacher, Hoffmann, Schmael, Schulte-Koerne,

& Noethen, 2007). This article focuses on the family elements, which influence children's reading abilities in China.

2 Literature

Spoken Chinese has only 1,200 syllables but over 5,000 commonly used morphemes (Zhou, 1978), and characters often have the same pronunciation. In order to understand the meaning of a character, readers have to distinguish among homophonic morphemes. Spoken Chinese has a relatively simple phonological structure. A syllable consists of an optional onset and a rhyme. The rhyme is formed by a vowel and an optional final consonant. Consonant clusters are rare. Chinese is also a tonal language. Every syllable in Mandarin Chinese is differentiated according to one of the four tones or voice inflections: high, rising, low then rising, or falling. Cantonese has nine tones. Mandarin contains only 403 syllables. Further, when differentiated by tone, there were about 1,200 distinct 'tone syllables' (Shu et al., 2005).

Reading and writing are important parts of children's language learning. In addition, visual perceptual processing and motor skills are factors that affect children's literacy. Family reading background is an important factor affecting the development of children's literacy. It was found that parents who are involved in children's reading and writing learning activities were related to the development of the children's early reading and writing skills (Sencechal & Lefovre, 2002). The contributions of family cultural background to four students in the first grade regarding reading skills were 10.3 % and 17.5 %, respectively by questionnaires and tests methods (Shu et al., 2002). Further, the family cultural background, as a whole, has a strong influence on first grade and fourth grade children's reading abilities. Although family reading background plays an important role on children's reading success, the role of other aspects of the family reading's background among children in different grades are not the same.

Chinese children have to learn about 2,500 visually complex characters. Most characters have seven to 12 strokes. About 72 % of the characters learned by children in primary school are semantic phonetic compounds (Shu, Chen, Anderson, Wu, & Xuan, 2003). Such a complex orthographic structure may cause many cognitive difficulties for Chinese children struggling to master the writing system (e.g., associating a specific character with speech and meaning). Studies on reading development in Chinese learners have shown that children's reading achievements are strongly related to children's phonological skills in understanding speech structure and manipulating phonemes and lexical tones (Ho & Bryant, 1997; McBride-Chang & Ho, 2000; Shu, Chen, Anderson, Wu, & Xuan, 2003; Siok & Fletcher, 2001). Some scholars investigated the relations between auditory and speech processing and reading development in Chinese school children. It was found that reading development in Chinese learners correlates strongly with children's ability in auditory and temporal processing. Deficits of information in auditory temporal, consonants, and vowel processing of Chinese syllables can also manifest in dyslexic children's brain responses to deviant stimuli in the oddball paradigm (Meng, Sai, Wang, Wang, Sha, & Zhou, 2005).

Due to the day-by-day increase in social information, children's development to the reading's demand is high increasingly. When children with reading disorders have not been identified and given treatment before the age of nine, their reading difficulty will remain in at least 74 % of the learners in high school (Shaywitz & Shaywitz, 1994). Previous studies indicated that if the students are identified and the reading problem is intervened early by special education or remedial education programs, as many as 70 % can overcome reading problems (Lyon et al., 2001).

In recent years, people have been paying more attention to the influence of the family culture environment on children's reading development. Family elements, such as the family cultural reading behavior, have a strong influence on language reading ability among Chinese children. The study found a high correlation between the number of family-owned books and the children's reading ability; the number of books available for children can largely explain the individual differences in the reading ability of children. In early days, parents read stories to their children and share pictures with their children, these can promote children's oral vocabulary, reading interest, and reading ability.

3 Method

Literature review is used in this article to describe reading problems among Chinese elementary school children and the role of family elements.

4 Results

In China, there are many research studies on the development of reading and reading problems focused on the inherent characteristics of reading (Table 1), such as the process of the form and meaning, understanding, and word recognition mechanisms (Meng, Zhou, & Kong, 2002). The study of the related factors of children's reading and writing abilities and the acquisition and development, such as the family reading background, is still very rare. Early studies of family influence on children's reading ability mainly focused on the family social background (e.g., family social economic situation, parents' occupation, and education).

Table 1. Reading Problems among Elementary School Children in China

Components of reading model	Findings
Cognitive components	Oral language ability & character/word recognition (Meng, Zhou, & Kong, 2002), understanding meaning (Meng, Zhou, & Kong 2003), lower phonological awareness (Shu et al., 2002).
Psychological components	Reading achievement (Shu et al., 2002).
Ecological	Different alphabetic writing system

Chinese studies have shown that the home-literacy environment significantly contributes to children's reading proficiency. The results of path analysis further demonstrate that parent-child literacy-related activities directly contribute to first-grader reading proficiency (about 6 or 7 years old). However, the four aspects of the home-literacy environment independently contribute to fourth-grader (about 10 or 11 years old) reading proficiency (Shu et al., 2002). On the other hand, family also can do something to correct children's reading problems.

In addition to children with reading problems, despite children receiving medical treatment, nowadays, many research studies focus on school education to prevent and correct children's reading problems. The problem is that children with reading problems cannot keep up with the normal rhythm of school teaching, and ordinary schools lack educational resources for children with reading problems. Most teachers also lack professional knowledge about how to help children with reading problems. Therefore, it is particularly important for the family environment to intervene and correct children's reading problems (see Table 2).

Table 2. Protective and risk family factors related to reading problems in China

Family elements	Findings
Risk factors	Family illiteracy (Meng, Zhou, & Kong, 2002); lack of parental education (Sun et al., 2013).
Protective factors	Reading time spending by parents' at home (Ke, 2013); positive family culture and learning environment (Sun et al., 2013).

Family is the place where children are born and grow; families can discover children's reading problems early. Parents are the first teachers for children and can be the earliest to intervene and correct the children's reading problems. Meanwhile, family environment is the important supplement to hospital treatment and school education. In view of the parents' familiarity and understanding of the children, the home environment is more targeted to intervene and correct reading problems of children, and perhaps the outcome is also the best.

Further, the study of children with reading difficulties is typically carried out in countries with a phonetic alphabet. Since Chinese characters are different from the alphabetic system, whether the Western conclusions are suitable for Chinese reading problems should be further studied. Research shows that Chinese reading disorders manifest in the following aspects: difficulty in recognition of form and sound as well as form and meaning, difficulty in reading fluency, difficulty in reading accuracy, and difficulty in comprehension.

Although a lot of research in Western countries has proved that the development of early phonological awareness will affect the children's future reading ability, there are inconsistent results when children use Chinese characters. The developmental process of reading ability in Chinese children and the role of phonological awareness do not seem so important. Chinese children's ability of early phonological awareness cannot effectively predict the future of reading ability in adulthood. Further, it was found there were three types of children's reading problems in China: (a) single word recognition obstacles accounting for 21.6 %, (b) phrase and sentence comprehension difficulties accounting for 8.1 %, and (c) mixed accounting for 70.3 % (Yang & Gong, 1998).

5 Discussion

This study investigates Chinese children's reading abilities and the relationship with related factors. Previous research showed that in China, the family reading background influences oral language ability and character recognition, while oral language ability influences meaning comprehension and character recognition (Meng, Zhou, & Kong, 2002). It was found that problems in reading ability among elementary students were mostly influenced by the parents and the learning environment.

6 Affiliations

M.Sc. Guangshu Gu
Institution: Wuhan Textile University
Address: 1 FangZhi Road, Wuhan, P. R. 430073, China
E-mail: guguangshu@163.com

M.A. Dian Sari Utami
Institution: Islamic University of Indonesia

Address: Jalan Kaliurang kms. 14.5, Sleman 55584 Yogyakarta, Indonesia
E-mail: dian.utami@uii.ac.id

7 References

Chen, D. (2007). *Progress and enlightenment of research on the intervention of children with dyslexia in the West* [in Chinese]. (Unpublished master thesis). Northeast Normal University, Changchun, China.

Ho, C. S.-H., & Bryant, P. (1997). Phonological skill are important in learning to read Chinese. *Developmental Psychology, 33*, 946–951.

Katusic, S. K., Colligan, R. C., Barbaresi, W. J., Schaid, D. J., & Jacobsen, S. J. (2001). Incidence of reading disability in a population-based birth cohort 1976–1982 Rochester, Minn. *Mayo Clin Proc, 76*, 1081–1092.

Ke, H. (2013). Reading is a necessary study pipeline in the new century [in Chinese]. *Reading research, 14*(4), 4–11.

Lyon, G. R., Fletcher, J. M., Shaywitz, S. E., Shaywitz, B. A., Torgesen, J. K., Wood, F. B., ... & Olson, R. (2001). Rethinking learning disabilities. In C. E. Finn, A. J. Rotherman, & C. R. Hokanson (Eds.), *Rethinking special education for a new century* (pp. 259–287). Washington DC: Thomas B. Fordham Foundation.

McBride-Chang, C., & Ho, C. S.-H. (2000). Developmental issues in Chinese children's character acquisition. *Journal of Educational Psychology, 92*, 50–55.

Meng, X. Z., Sai, X. G., Wang, C. X., Wang, J., Sha, S. Y., & Zhou, X. L. (2005). Auditory and speech processing and reading development in Chinese school children: Behavioral and ERP evidence. *Dyslexia, 11*, 292–310.

Meng, X. Z., Zhou, X. L., & Kong, R. F. (2002). Chinese literacy and its related factors [in Chinese]. *Psychological Science, 25*(5), 544–572.

Meng, X. Z., Zhou, X. L., & Kong, R. F. (2003). Structure model of Chinese reading abilities and its correlate [in Chinese]. *Psychological Development and Education, 1*, 37–43.

Schumacher, J., Hoffmann, P., Schmael, C., Schulte-Koerne, G., & Noethen, M. M. (2007). Genetics of dyslexia: the evolving landscape. *Journal of Medical Genetics, 44*(5), 289–297.

Sencechal, M., & Lefovre, J. (2002). Parental involvement in the development of children's reading skill: a five-year longitudinal study. *Child Development, 2*, 445–460.

Shaywitz, B. A., & Shaywitz, S. E. (1994). Learning disabilities and attention disorders. In K. Swaiman (Ed.), *Principles of pediatric neurology*. (pp. 1119–1151). St. Louis: Mosby.

Shu, H., Chen, X., Anderson, R. C., Wu, N., & Xuan, Y. (2003). Properties of school Chinese: Implications for learning to read. *Child Development, 74*, 27–47.

Shu, H., Li, W. L., Ku, Y. M., Anderson, R. C., Wu, X. C., Zhang, H. C., & Xuan, Y. (2002). The role of family and cultural background in children's reading development [in Chinese]. *Psychological Science, 25*(2), 136–139.

Shu, H., Meng, X. Z., Chen, X., Luan, H., & Cao, F. (2005). The subtypes of developmental dyslexia in Chinese: Evidence from Three Cases. *Dyslexia, 11*, 311–329.

Siok, W. T., & Fletcher, P. (2001). The role of phonological awareness and visual-orthographic skills in Chinese reading acquisition. *Developmental Psychology, 37*, 886–899.

Sun, Z., Zou, L., Zhang, J., Mo, S., Shao, S., Zhong, R., … & Song, R. (2013). Prevalence and associated risk factors of dyslexic children in a middle-sized city of China: A cross-sectional study. *PLoS ONE 8*(2), e56688.

Yang, Z., Gong, Y., & Li, X. (1998). Clinical features and subtyping of Chinese children with reading disorder [in Chinese]. *Chinese Journal of Clinical Psychology, 6*(3), 136–139.

Zhou, Y. G. (1978). To what degree are the 'phonetics' of present-day Chinese characters still phonetic? [in Chinese]. *Zhongguo Yuwen (Studies of the Chinese Language), 146*, 172–177.

Chapter 2
Traumatic Experiences

Zarina Akbar[1, 2] & Evelin Witruk[1]

[1] University of Leipzig, Germany

[2]Jakarta State University, Indonesia

Coping Strategies and Disaster Experience Predict Post-traumatic Growth Survivors of Disaster in Yogyakarta Province Indonesia

Abstract. This study examines association among coping strategies, disaster experience, and post-traumatic growth in disaster survivors. The samples consists of 100 survivors, female 65 ($M = 3.98$; $SD = 0.35$) and male 35 ($M = 3.78$; $SD = 0.31$) affected by natural disaster, earthquake 2006 in Bantul district ($n= 50$) and volcano eruption 2010 in Cangkringan Sleman district ($n= 50$) in Yogyakarta Province, Indonesia. Data were collected several years after disasters in 2013. The measurement instrument used for data collection has subscales on coping, disaster experience, and post-traumatic growth level. Coping aspects were divided into approach and avoidance coping. Possible predictors to post-traumatic growth were examined by regression analyses. The result showed that coping and disaster experience are siginificant predictors to post-traumatic growth. Implications for this research further highlight the need for addressing approach coping and disaster experience, which are more important than the nature of traumatic event in rehabilitation programme for disaster survivors.

Keywords: coping strategies, disaster experience, post-traumatic growth, disaster survivors.

1 Introduction

Over the past decade, the world has seen an increase in large-scale traumas, catastrophes, and natural disaters. According to the United Nations (UN), natural disasters are increasing in frequency and severity around the globe (Jacobs, Leach, & Gerstein, 2011). The rising population is one of the contributors to the negative impacts of natural disasters because death tolls and devastations are greater in areas with a more dense population. Indonesia is a hazard-prone country as it is situated at the meeting point of three active plates in the world: the Indo-Australian plate in the south, the Euro-Asian plate in the north, and the Pacific plate in the east. The three plates are moving and thrusting towards one another in such a way making the area prone to natural disasters such as volcanoes, earthquakes, tsunamis. The movement of the plates also causes the area to become a tectonically and volcanically active region (National Agency for Disaster Management, 2010). Thus, natural disasters occur almost every year in Indonesia, some of which hit Yogyakarta Province in May 2006 and October 2010 when two massive disasters, a catastrophic tectonic earthquake and a volcanic eruption of Mount Merapi, caused many casualties and extensive property damage.

Natural disasters often cause a number of psychological distresses, but post-traumatic stress disorder particulary happens when there are many casualties in the disaster. The first reaction of the individual to disasters varies ranging from a state of shock, fear, sadness, and anger, which may be leading to a denial to the catastrophic events that have just occurred. The individual's ability to control his life decreased and a lot of predictable and real things threatened by the arrival of this unexpected disaster. The disaster did not only give negative impacts, but also positive ones. These positive changes and experiences are called *Post-traumatic growth* or PTG (Karanci & Acarturk, 2005). PTG is the process of getting and maintaining perceived positive outcomes from a traumatic experience (Tedeschi, Park, & Calhoun, 1998). Many terms including found meaning, benefit finding, post-traumatic or stress-related growth, perceived benefits and self-transformation have been used to capture the experience of positive change or growth. In particular terms of post-traumatic growth has been used in reference to 'a sense' that personal growth resulted from a challenging life experience. Taking into consideration cross-cultural aspects, a contribution shall be thus made to explore the long-term consequences of natural disasters in Yogyakarta Province Indonesia. Why did some survivors reach PTG? How did the people dealing with the devastating disasters? What impacts did disaster experience influence PTG? What is the relationship between coping and disaster experience to PTG?

2 Literature

2.1 Post-traumatic growth (PTG)

Tedeschi and Calhoun (1996) have noted that traumatic events that confront the individual may become a challenge of how to make the experience manageable, comprehensible, and meaningful. Successful adaptation requires effective negotiation of these psychological tasks, which in turn can provide the base for positive individual and interpersonal changes. Also Calhoun, Cann, Tedeschi, and McMillan (2000) defined *post-traumatic growth* as "positive change that an individual experiences as a result of the struggle with a traumatic event".

2.2 Coping strategies

Coping is different from the automatic habitual responses that are perhaps measured more appropriately by psychological scale measurement. Coping is a process by which an individual manages the demands and emotions generated by that which is appraised as stressful (Lazarus, 1999). Strategies include appraisals of a stressful event and bestowing the situation with meaning, as opposed to the global meaning assessed when measuring levels of PTG (Folkman & Moskowitz, 2000). The process involves appraisals as to whether a situation is a threat, a challenge, or a loss, and perceptions of what can be done to alter the situation or minimise the threat (Lazarus & Folkman, 1984).

2.3 Disaster experience

Disaster experiences indicated positive changes in aftermath. Research found that 34 % of surveyed survivors indicated positive changes in the aftermath of the devastating earthquake and tsunami disaster in Phuket Thailand (So-kum, 2006). A study also found 90–95 % of surveyed tornado survivors in the United States reported positive benefits at 4–6 weeks as well as 3 years following the disaster (McMillen, North, & Smith, 2000). It found that tsunami disaster-related traumatic exposure such as personal injuries, loss of or damage to property, and relocation to temporary shelters were related to positive post disaster adjustment (MaFarland & Alvaro, 2000; Park, Cohen, & Murch, 1996).

3 Method

This study examines coping and disaster experience variables in relation to levels of post-traumatic growth among disaster survivors in Yogyakarta Province Indonesia.

3.1 Participants and procedure

The participant consists of 100 survivors of affectedness natural disaster earthquake in Bantul district and volcano eruption in Cangkringan Sleman district in Yogyakarta Province, Indonesia. For the period July-September 2013, a process data was collected. The participants were personally approached, given information about the purpose of the research, and invited to participate. Confidentially of information and its restricted use for research only were assured. The participant was 35 % male and 65 % female. Fifty percent of study participants were survivors from the Bantul's earthquake 2006 and 50 % of the Merapi Eruption 2010. The educational background of the participants was no school 6 %, elementary school 36 %, junior high school 25 %, senior high school 32 %, and university 1 %. The marital status was single 12 %, married 82 %, and widow/widower 6 %. All participants were Muslim. Participant's occupation was labourer 19 %, teacher 2 %, housewife 31 %, and others 48 %.

3.2 Measures

The measurement instruments used for data collection had subscales on coping, disaster experience, and post-traumatic growth level.

Coping. Coping was assessed with a 26 items scale. These items were derived from 28 items of the Brief COPE Scale (Carver, 1997) and were selected based on their high factor loadings. Substance usewere not examined in this research because of low factor loading analysis and related to culture bias. The Brief COPE has adequate internal reliability (Carver, 1997). Finally, coping aspects divided into approach and avoidance coping.

Disaster Experience. The disaster experience scale consists of 10 items. This scale has been modified from earthquake experience scale by Karanci and Acarturk (2005). It contained ten items related to the earthquake experience and impact (e.g., previous earthquake experience, whether the respondents had thoughts about family membes/relatives dying during the quake, whether they saw a dead or injured person after the earthquake).

Post-Traumatic Growth. Post-traumatic growth was assessed with the 21-items Post-Traumatic Growth Inventory (Tedeschi & Calhoun, 1996) that include aspects of perceptions of growth in relating to others, new possibilities, personal strength, spiritual change, and appreciation of life. The Post-Traumatic Growth Inventory was developed to assess growth-related changes experienced by traumatized individuals. The 21-item scale yields a total score and five subscale scores: New Possibilities, Relating to Others, Personal Strength, Spiritual Change, and Appreciation of Life.

4 Results

Possible predictors to post-traumatic growth were examined by regression analyses in this research. A regression analysis in Table 1 showed that coping and disaster experience are significantly predicted post-traumatic growth ($F = 6.59$; $p < 0.05$). Disaster experience significantly predicted post-traumatic growth ($\beta = .41$). Among those aspects significantly predicted 17.10 % to post-traumatic growth.

Table 1. Regression Analysis Result

ANOVA[a]

Model	Sum of Squares	df	Mean Square	F	Sig.
Regression	2.05	3	.68		
Residual	9.94	96	.10	6.59	.00[b]
Total	11.99	99			

Note a. Dependent Variable: PTG; b. Predictors, Coping Avo, Coping App, Disaster Experience

5 Discussion

Based on the statistical regression analysis, it was found that coping and disaster experience confirmed significant relationship with post-traumatic growth scores. Psychological research has long held an interest in identifying coping abilities that promote better adjustment in the aftermath of trauma. Many coping theories assume that survivors of trauma engage in a cognitive process of certaining meaning in relation to their experience in order to successfully cope with it (Folkman, 2008; Tedeschi & Calhoun, 1996). Coping is often referred to in terms of strategies, styles, resources, approaches, and skills. These terms may differ conceptually and the coping used by one individual to another is also different. Alternatively, other researchers use the term strategy and advocate a contextual response, whereby coping is viewed as being flexible across situations and over time (Skinner, Edge, Altman, & Sherwood, 2003).

Moreover, current coping theories contend that the effectiveness of any given strategy is dependent on the context of the traumatic incident (Schulz & Mohamed, 2004). According to this view, any particular strategy employed by the person to deal with the trauma can be either adaptive or maladaptive, depending on the circumstance. For example, Whealin, Ruzek, and Southwick (2008) reviewed a number of studies that have referred to adaptive and maladaptive coping influences, and other studies have differentiated coping strategies by using terms such as, functional or dysfunctional, transformation or regressive coping, also Sharkansky, King, Wolfe, Erickson, and Stokes (2000) who examined the relationship between approach or avoidance coping strategies on the psychological ill-health in active military personnel. The research result found that approach coping are recognized to post-traumatic growth in disaster survivors. Approach coping relates to direct attempts at problem-solving activities to relieve the source of psychological distress and relieve through positive reframing, and an optimistic outlook. It was found that approach coping was more effective than avoidance coping in regard to post-traumatic growth of the disaster survivors.

Disaster experience influenced post-traumatic growth in survivors. Disaster experiences contained severity of impact and perceived life threat aspects. It found that post-traumatic growth was significantly correlated with the perceived of severity of impact and perceived of life threat (Karanci & Acarturk, 2005). Also Calhoun and Tedeschi (1998) stated that being severely exposed to a trauma shatters the assumptions of the survivor, and this enables the person to change them. Those with disaster experiences got a more severe exposure, which in turn may have enabled them to experience higher levels of growth. Implications for this research offer further highlight the needs for addressing coping especially approach coping and disaster experience, which are more important than the nature of traumatic event in rehabilitation program for disaster survivors. Further research may need to analyse other possibilities also related psychological aspects to post-traumatic growth in difficult life experiences, such as social support and spirituality in disaster survivors.

6 Affiliations

Dr. Zarina Akbar
Institution: Department of Educational and Rehabilitation Psychology, University of Leipzig and Department of Psychology, Jakarta State University
Address: Neumarkt 9–19, 04109 Leipzig, Germany and Rawamangun Muka Street, East Jakarta, 13220 Indonesia.
E-mail: zarinaakbar@yahoo.com

Prof. Dr. Evelin Witruk
Institution: University of Leipzig, Educational and Rehabilitation Psychology
Address: Neumarkt 9–19, 04109 Leipzig, Germany
E-mail:witruk@ uni-leipzig.de

7 References

Calhoun, L. G., & Tedeschi, R. G. (1998). Post-traumatic growth: future directions. In Tedeschi, R. G., Park, C. L., & Calhoun, L. G. (Eds). *Post-traumatic growth: Positive change in the aftermath of crisis* (pp. 215–238). Mahwah, NJ: Lawrence Erlbaum.

Calhoun, L. G., Cann, A., Tedeschi, R. G., & McMillan, J. (2000). A correlational test of the relationship between post-traumatic growth, religion, and cognitive processing. *Journal of Traumatic Stress, 13*(3), 521–527.

Carver, C. S. (1997). You want to measure coping but your protocol's too long: Consider the Brief COPE. *International Journal of Behavioral Medicine, 4*(1), 92–100.

Folkman, S. (2008). The case for positive emotions in the stress process. *Anxiety, Stress, and Coping, 21*, 3–14.

Folkman, S., & Moskowitz, J. T. (2000). Positive affect and the other side of coping. *American Psychologist, 55*(6), 647–654.

Jacobs, S. C., Leach, M. M., & Gerstein, L. H. (2011). Introduction and overview: Counselling psychologists' roles, training, and research contributions to large-scale disasters. *The Counselling Psychologist, 39*(8), 1070–1086.

Karanci, N & Acarturk, C. (2005). Post-traumatic growth among Marmara Earthquake survivors involved in disaster preparedness as volunteers. *Traumatology, 11*(4), 307–323.

Lazarus, R. S., & Folkman, S. (1984). *Stress, appraisal, and coping.* New York: Springer.

Lazarus, R. S. (1999). *Stress and emotion: A new synthesis.* New York: Springer.

MaFarland, C., & Alvaro, C. (2000). The impact of motivation on temporal comparisons: Coping with traumatic events by perceiving personal growth. *Journal of Personality and Social Psychology, 79*, 327–343.

McMillen, J. C., North, C. S., & Smith, E. M. (2000). What parts of PTSD are normal: intrusion, avoidance, or arousal? Data from the Northridge, California earthquake. *Journal of Trauma and Stress, 13*, 57–75.

National Agency for Disaster Management. (2010). *National Disaster Management Plan 2010–2014.* Jakarta: Indonesia.

Park, C. L., Cohen, L. H., & Murch, R. L. (1996). Assessment and prediction of stress-related growth. *Journal of Personality, 64*(1), 71–105.

Tedeschi, R. G., & Calhoun, L. G. (1996). The Post-traumatic growth Inventory: Measuring the Positive Legacy of Trauma. *Journal of Traumatic Stress, 9*(3), 455–471.

Tedeschi, R. G., Park, C. L., & Calhoun, L. G. (1998). *Post-traumatic growth: Positive Changes in the Aftermath of Crisis.* London: LEA, Inc.

Schulz, U., & Mohamed, N. E. (2004). Turning the tide: Benefit finding after cancer surgery. *Social Science & Medicine, 59*, 653–662.

Sharkansky, E. J., King, D. W., King, L. A., Wolfe, J., Erickson, D. J., & Stokes, L. R. (2000). Coping with Gulf War combat stress: Mediating and moderating effects. *Journal of Abnormal Psychology, 109*, 188–197.

Skinner, E. A., Edge, K., Altman, J., & Sherwood, H. (2003). Searching for the structure of coping: A review and critie of category systems for classifying ways of coping. *Psychological Bulletin, 129*, 216–269.

So-kum Tang, C. (2006). Positive and negative postdisaster psychological adjustment among adult survivors of the Southeast Asian earthquake-tsunami. *Journal of Psychosomatic Research, 61,* 699–705.

Whealin, J. M., Ruzek, J., I., & Southwick, S. (2008). Cognitive behavioural theory and preparation for professionals at risk for trauma exposure. *Trauma, Violence, & Abuse,* 9, 100–111.

Gunendra R. K. Dissanayake

University of Peradeniya, Peradeniya, Sri Lanka

Trauma Never Ending: The Impact of Different Forms of Intimate Partner Violence (IPV) on Women's Psychological Well-being

Abstract. The objective of the present study was to investigate the relative contribution of psychological, sexual and physical abuse to the development of psychological distress among women experiencing intimate partner violence (IPV) in Sri Lanka. A sample of 200 help seeking ever-partnered women recruited from ten women help centers from five districts in Sri Lanka completed a several structured questionnaires measuring different types of abuse, and psychological symptoms, as well as demographic variables. Findings show that there was a high degree of overlap between the different types of abuse, in 79 % of cases physical, sexual and psychological abuse occurring together. Multiple regression analyses revealed that psychological abuse was a stronger predictor of psychological distress, memory problems, problems in general functioning, and attempts of suicide than physical and sexual abuse, even though psychological, physical, and sexual abuse experiences were highly correlated. Psychological abuse contributed uniquely to the prediction of psychological distress, even after controlling for the effects of physical and sexual abuse. Results highlight the importance of examining the effects of less visible forms of IPV like psychological abuse, independent of physical forms of abuse, to understand its impact on victims.

Keywords: intimate partner violence, mental health, psychological distress, Sri Lanka

1 Introduction

Intimate partner violence (IPV) causes short- and long-term negative consequences, both physical and psychological (e.g., Campbell, 2002; Ellsberg et al., 2008). In Sri Lanka, empirical studies on the subject only emerged in the past two decades, and these estimate the prevalence of IPV in the range of 18 % to 72 % in different populations and age groups (Deraniyagala, 1992; Jayathilake, Poudel, Yasuoka, Jayatilleke, & Jimba., 2010; Moonesinghe, 2002). Despite the high prevalence of IPV against women in Sri Lanka, there is limited literature on the consequences of IPV against women. Hence, the objectives of the present study is to investigate the relative contribution of main forms of IPV: psychological, sexual and physical abuse to the development of physical and mental health consequences among women in Sri Lanka.

2 Literature

Despite increasingly well-documented literature on the prevalence of intimate partner violence and its impact on women's psychological well-being (Campbell, 2002; World Health Organization[WHO] 2013), many gaps still exit in the scientific literature. With noted exceptions (Campbell, Torres, & Ryan, 1995; Garcia-Moreno et al., 2006; Wagner, Mongan, Hamrick, & Hendrick, 1995), most past studies that address the health effects of IPV measured physical assaults alone without considering the long-term psychological or/and sexual abuse characteristic of violent relationships. However, it is possible that not only physical abuse, but the other forms of IPV also exert a detrimental impact on women's psychological well-being. The negative consequences of IPV might vary according to the type of violence victims experience: physical, psychological, or sexual (Arias & Pape, 1999; Herbert, Silver, & Ellard, 1991; Vitanza, Vogel & Marshall, 1995). Hence, the examination of the effects of other forms of IPV also would be an important research endeavor.

3 Method

3.1 Sample

A sample of 18–49 year old 200 help seeking ever-partnered women were randomly selected from ten women help centers from five districts in Sri Lanka to take part in this cross-sectional survey.

3.2 Procedure

The study was conducted by following the WHO's ethical guidelines for conducting research into "violence against women" (WHO, 2001). Ethical approval was obtained from the University of Colombo, Sri Lanka. All participants signed an informed consent form prior to the administration of the questionnaire. Data collection was carried out by trained research assistants using a structured questionnaire.

3.3 Measurement

Psychological Maltreatment of Women Inventory (PMWI). The PMWI was developed by Tolman (1989). The 58 item PMWI-long form was adapted and validated in the Sri Lankan context. The instrument consists of two factor-derived subscales that measure dominance and isolation and emotional, verbal, monitoring abuse. The scale is a self-report measure, and each item is rated on a 5-point frequency scale. In the current study, the Cronbach's Alpha for this scale was .94, and for the the subscales, the internal alpha coefficient of reliability were .88 and .89 respectively.

Physical and sexual abuse questionnaire. The adapted and validated version of the women's health and life events questionnaire developed by the WHO (Garcia- Moreno, Heise & Ellsberg, 2001) was used to examine physical and sexual abuse among the participants. Experience of physical abuse was assessed using 6 items and sexual abuse was assessed using 3 items. If a respondent answered "yes" to 1 or more of the items of physical or sexual abuse, she was considered to be physically or sexually abused.

The Questionnaire on General Health (QGH). Mental health was assessed using a self-reporting questionnaire of 20 questions (SRQ-20), developed by WHO as a screening tool for psychological distress. The alpha coefficients for the QGH are 0.90.

4 Results

Characteristics of the Female Respondents. The mean age of the respondents was 34.22 (SD=7.50), ages ranging from 20–49. The majority of the respondents (78 %) were married, 1.5 % were living together, 19 % were separated and another 1.5 % were divorced at the time of the study. The majority of the women (88 %) had received secondary level of education, whereas both the primary and tertiary levels consisted of 6 % each.

Prevalence of IPV. On average, 96 % of the women experienced at least one or more incidents of physical abuse in the preceding six months of the survey (M= 16.89, SD= 6.14). The total sample (100 %) reported experiencing some degree of psychological abuse, whereas 81 % of the respondents reported that they have been subjected to some degree of sexual abuse by their husbands/partners. In most instances (79 %), psychological abuse occurred together with physical and sexual abuse suggesting that these multiple types of abuse tend to co-occur. Figure 1 presents the occurrence of IPV by type.

Figure 1. Occurrence of IPV by type

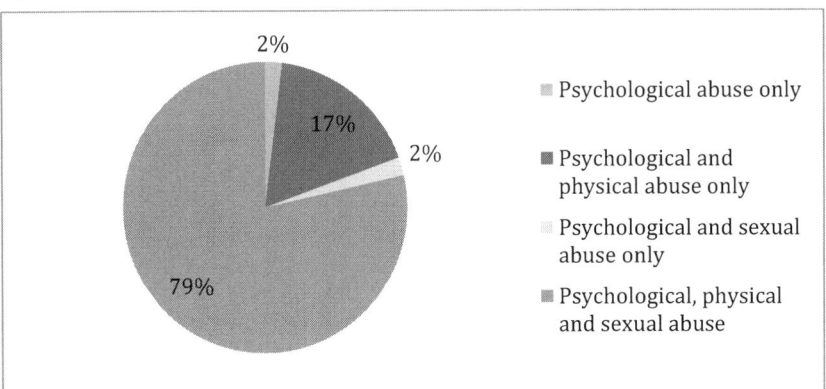

Psychological Distress. The mean score of psychological distress for women who experienced abuse (*M*= 12.47) was significantly higher than that of non-abused women (*M*= 5.4). Ever experiencing psychological IPV was associated with a significant increase in risk of developing the following conditions: problems in functioning, memory problems, headaches, feeling nervous and tensed, digestion difficulties, dysfunction in daily activities, feeling tired all the time, feeling as a worthless person, thoughts of ending life, and attempts in committing suicide (*p* =.05). Sexual IPV was also significantly associated with the following: feeling nervous and tensed, trouble thinking clearly, difficulty in making decisions, feeling tired all the time, and thoughts of ending life (*p* =.05). Physical IPV was significantly associated with the following a few conditions when compared with the other two forms of IPV (i.e., feeling unhappy and digestion difficulties (*p* =.05)). When compared with other two forms, Psychological abuse contributed uniquely to the prediction of psychological distress, even after controlling for the effects of physical and sexual abuse. Further, psychological abuse was a stronger predictor of memory problems, problems in general functioning, and attempts of suicide than physical and sexual abuse.

Table 1. Contribution of Different Forms of IPV to Psychological Distress

	Unstandardized Coefficients		Standardized Coefficients		
Model	B	Std. Error	Beta	t	Sig.
1 (Constant)	12.801	1.232		10.393	.000
Physical abuse total score	-.061	.070	-.068	-.880	.380
Sexual abuse total score	.086	.110	.060	.781	.436
2 (Constant)	4.501	2.320		1.940	.054
Physical abuse total score	-.099	.068	-.110	-1.469	.143
Sexual abuse total score	-.097	.114	-.068	-.845	.399
Psychological abuse total score	.051	.012	.321	4.159	.000

a. Dependent Variable: Distress total score

5 Conclusion

The current study was an attempt to determine the relative contribution of psychological, sexual and physical abuse to the development of psychological problems among women experiencing IPV in Sri Lanka. Findings showed that psychological, physical, and sexual abuse were highly correlated. It was not possible in this study to separate psychologically abused women from physically or sexually abused women, because all women reported experiencing at least one form of each. However, the results demonstrated that the occurrence of physical or sexual abuse in the absence of psychological abuse was rare, indicating that psychological abuse present in the intimate relationship signals that physical or sexual abuse, or both may follow later. This finding is consistent with the previous findings which have shown that psychological abuse can precede, follow, or occur concurrently with other forms of violence (Stets, 1991; Tolman, 1992). It further implies the importance of screening for psychological abuse in early in the relationship as a preventive measure of the occurrence of physical and sexual abuse later in the relationship.

Results indicated that experiencing physical, psychological and sexual abuse was associated with psychological symptoms such as problems in functioning, memory problems, digestion difficulties, feeling tired all the time, feeling as a worthless person, thoughts of ending life, and attempts in committing suicide. Similarly, past research also have shown that women who reported IPA in the past year to be at increased risk for a number of serious emotional and physical health concerns, including depression, anxiety, sleep problems, suicidal ideation, and disabilities in functioning (Hathaway et al., 2000). However, multiple regression analyses revealed that psychological abuse was a stronger predictor of psychological distress, memory problems, problems in general functioning, and attempts of suicide than physical and sexual abuse, even though psychological, physical, and sexual abuse experiences were highly correlated. Psychological abuse contributed uniquely to the prediction of psychological distress, even after controlling for the effects of physical and sexual abuse. The higher the level of partner psychological abuse, the more the women experienced limitationson physical activities and role functioning, cognitive impairment, and negative perceptions of health. These findings not only confirm previous research showing that psychological abuse has a unique impact on victim outcomes over and above physical and sexual abuse (Arias & Pape, 1999; Herbert et al., 1991; Vitanza et al., 1995) but also suggest that psychological abuse has its greatest impact on health through decrements in health status (e.g., perceptions of health, functional impairment). Strong relationship between psychological abuse and health status, which in part reflects subjective beliefs about health, is consistent with theories of somatization and suggests that somatic-based complaints may be an expression of victim psychological distress. Results highlight the importance of examining the effects of less visible forms of IPV like psychological abuse, independent of physical forms of abuse, to understand its impact on victims.

In sum, psychological victimization is at least as detrimental and may be more detrimental than physical abuse for individual psychopathology symptoms among abused women. To reduce the range of health consequences associated with IPV, clinicians should screen for psychological forms of IPV as well as physical and sexual IPV.

6 Affiliation

M. Phil. Gunendra R. K. Dissanayake
Institution: University of Peradeniya, Peradeniya, Sri-Lanka
Address: Department of Philosophy and Psychology, University of Peradeniya, Sri Lanka.
E-mail: asabhakeeragala@gmail.com

7 References

Arias, I., & Pape, K. T. (1999). Psychological abuse: Implications for adjustment and commitment to leave violent partners. *Violence and Victims, 14*(1), 55–67.

Campbell, J. (2002). Health consequences of intimate partner violence. *The Lancet, 359*, 1331–1336.

Campbell J., Torres, S., Ryan, J. et al. (1999). Physical and nonphysical partner abuse and other risk factors for low birth weight among full term and preterm babies: a multiethnic case-control study. *Am J Epidemiol, 150*, 714–726.

Deraniyagala, S. (1992). *Domestic Violence.* Colombo: Women in Need.

Hathaway, J. E., Mucci, L. A., Silverman, J. G., Brooks, D. R., Mathews, R., & Pavlos, C. A. (2000). Health status and health care use of Massachusetts women reporting partner abuse. *American Journal of Preventive Medicine, 19*, 302–307.

Herbert, T. B., Silver, R. C., & Ellard, J. H. (1991). Coping with an abusive relationship: How and why do women stay? *Journal of Marriage and the Family, 53*, 311–325.

Ellsberg, M., Jansen, H. A. F. M., Heise, L., Watts, C. H., Garcia-Moreno, C., & the WHO Multi-country Study on Women's Health and Domestic Violence against Women Study Team (2008). Intimate partner violence and women's physical and mental health in the WHO Multi-country Study on Women's Health and Domestic Violence: An observational study. *Lancet, 371*, 1165–1172.

Jayatilleke, A. C., Poudel, K. C., Yasuoka, J., Jayatilleke, A. U., Jimba, M. (2010). Intimate Partner Violence in Sri Lanka. *Biosci Trends, 4*(3), 90–5.

Garcia-Moreno, C., Jansen, H. A. F. M., Ellsberg, M., Heise, L., & Watts, C. H. (2006). Prevalence of intimate partner violence: Findings from the WHO multi-country study on women's health and domestic violence. *Lancet, 368*, 1260–1269.

Garcia-Moreno, C., Heise, L., & Ellsberg, M. (2001). *WHO Multi-country Study on Women's Health and Life Events* (Final Core Questionnaire Version 10.0). Geneva: World Health Organization.

Moonesinghe, L. N., Rajapaksa, L. C., Samarasinghe, G. (2004). Development of a screening instrument to detect physical abuse and its use in a cohort of pregnant women in Sri Lanka. *Asia Pac J Public Health, 16*, 138–144.

Stets, J. E. (1991). Psychological Aggression in dating relationships: The role of interpersonal control. *Journal of Family Violence, 6*, 97–114.

Tolman, R. M. (1989). The development of a measure of psychological maltreatment of women by their male partners. *Violence and Victims, 4*, 159–177.

Tolman, R. M. (1992). Psychological abuse of women. In R. T., Ammerman. (Ed), *Assessment of Family Violence* (pp. 291–310). Oxford: John Wiley & Sons.

Vitanza, S., Vogel, L. C. M., & Marshall, L. L. (1995). Distress and symptoms of post-traumatic stress disorder in abused women. *Violence and Victims, 10*, 23–34.

Wagner P. J., Mongan, P., Hamrick, D., Hendrick, K. L. (1995). Experience of abuse in primary care patients: Racial and rural differences. *Arch Fam Med, 4*, 956–962.

World Health Organization (WHO). (2013). *Global and regional estimates of violence against women: Prevalence and health effects of intimate partner violence and non-partner sexual violence*. Geneva: Author.

World Health Organization (WHO). (2001). *Putting Women First: Ethical and Safety Recommendations for Research on Domestic Violence Against Women* Geneva: Author.

Asanka Bulathwatta

University of Leipzig, Germany

Trauma among University Students in Sri Lanka and Germany

Abstract. University education is an important stage of students' academic life. Therefore, all students need to develop their competencies to attain the goal of passing examinations and also to developing their wisdom related to scientific knowledge they gathered through their academic life. Life in universities is a critical period for individuals as it is a foot step to acquiring the emotional and social qualities in their social life. There are many adolescents who have been affected by traumatic events during their life span but have not been identified or treated. More specifically, there are numerous burning issues within first year university students namely, ragging done by seniors to juniors, bullying, invalidation and issues related to attitudes changes and orientation. Those factors can be traumatic for both their academic and day to day life style. Identifying the students who are with emotional damages and their resiliency afterwards the traumas and effective rehabilitation from the traumatic events is immensely needed in order to facilitate university students for their academic achievements and social life within the University education. This Study tries to figure out the role of emotional intelligence for developing coping strategies among adolescents who face traumatic events. Late adolescence students enrolled at University (Bachelor students) will be selected as sample. The study is to be conducted in a cross cultural manner comparing 400 students each from Germany and Sri Lanka.

Keywords: emotional intelligence, trauma, ragging, bullying, coping.

1 Introduction

This study explores the role of Emotional Intelligence in developing coping strategies among university students that have faced traumatic events. The sample is made up of university students. There are a number of commonly identified traumas among university students in Sri Lanka. Most of the traumatic events that have an effect on university students' careers can be attributed to their personal life as well as to societal events. It is noteworthy that some cultural differences exist when compared to European students and related traumatic events. While European society is based on an individualistic lifestyle resulting in traumas derived from individual events, Asian students have to deal with traumatic events on a more collective level as life patterns are characterized by the importance of group identity.

In general, late adolescents enter universities in a critical state of personal development combined with the challenging task of acquiring the emotional and social skills necessary for their future lives. There are many adolescents who have been affected by traumatic events during their lifetime yet have not been identified or treated accordingly. More specifically, there are numerous issues among first year university students, namely, ragging done by seniors to the juniors, bullying, cyber bullying, invalidation and issues relates to attitudes changes and orientation.

During their university education, all students need to develop competencies necessary to pass examinations and to develop their skill sets in regard to up-to-date scientific knowledge. In a friendly and effective academic environment students are enabled to regulate their emotions and to acquire emotional knowledge they can apply in their everyday life. Relationships developed among university students and their peers can also be considered an important factor in students' academic life. For instance, a student in a stable emotional state is able to develop appropriate relations with his or her peers. Moreover, rehabilitated students overcoming traumatic experiences using their emotional knowledge are a valuable resource and future investment for a country.

Rehabilitation efforts based on the concept of Emotional Intelligence[1] as utilized in this study, is a comparatively new application. However, this approach allows for positive outcomes with long-term benefits among adolescents facing trauma. Using a rehabilitation approach that is based on Emotional Intelligence further enables these victims to develop protective mechanisms for future traumatic events.

Participants are being enabled to develop a state of mental well-being that is grounded in and characterized by cognitive behaviours and emotions beneficial within the rehabilitation process. Thus, this form of rehabilitation based on coping strategies using Emotional Intelligence empowers participants to develop their reasoning abilities and further strengthens emotional attention, emotional clarity and emotional repair capacity necessary to cope with traumatic event as pointed out by Mayer in 2008.

There are number of possible traumas among university students. As was mentioned earlier, it can come in many forms such as bullying, hazing, invalidations based on subjective matters such as personality, and personal background, cast system, physical conditions, cultural discriminations, pledging and fagging. Ragging is the most common traumatic event among university students which

---

1   The ability to carry out accurate reasoning about emotions and the ability to use emotions and emotional knowledge to enhance thoughts (Mayer, 2000)

senior students trying to control juniors. In terms of concept familiarization junior students into the University context. In many cases ragging has an effect on our emotions and can also lead to physical handicaps in certain situations. The history of ragging underwent a massive transformation after World-War I. Soldiers returning from war to the college started to perform military sub-cultural practices in the academic setting. This is how ragging emerged in the educational context. The first ragging-related death has been reported in 1873 at Cornell University in the US. In general ragging is a technique which is used to persuade an individual that he/she fails as an individual and can only succeed as a team.

It is noteworthy that ragging no longer exists in its most brutal forms in the places it originated in but experiences a rapid proliferation in developing countries. At present, Sri Lanka is the most affected country in the world. A telling example is that of a 22 year old Faculty of Agriculture student at the University of Peradeniya who ended up paralyzed as a result of jumping from the second floor of a hostel to escape the physical ragging by university seniors. Even more disturbing, this student later committed suicide. This story is not an isolated case: Another horrendous ragging incident occurred in the same year when 21 year old S. Varapragash, an Engineering student of the University of Peradeniya, died from a kidney failure following severe ragging by senior students. Only at this point, the general public and government officials realized the seriousness of this issue at local universities.

Due to the fact that proper trauma rehabilitation is needed in order to facilitate university students for better coping within their traumatic events faced in the University life. In my study I will be trying to identify whether Emotional Intelligence has an effect on developing trauma coping strategies among university students.

2 Literature

There are a number of studies that addressed Post Traumatic Stress Disorder (PTSD) and the rehabilitation process. Symptoms of traumatic events can emerge three or four months after the traumatic event. The level of symptoms among individuals is determined by the traumatic events witnessed and the initial exposure to these events as well as by pre-existing demographic characteristics, the occurrence of major life stressors, the availability of social support, and the types of coping strategies used to cope with disaster related stress. There is a potential utility of those factors to organize thinking and predicting the emergence and persistence of PTSD symptoms among victims (La Greca, Silverman, Vernberg, & Prinstein, 1996). Silver and colleagues in 2006 have discovered how man-made attacks can have an effect on developing post-traumatic stress across the United States and conducted immediately after September 11th terrorist attack to the World Trade Center and the pentagon. In 2001, those scholars have identified variables that predicted who was most likely to suffer greater long-term psychological consequences of traumatic events. They suggest that the prior mental health history, prior life traumas, as well as the significant role of subsequent stressors, are significant factors in explaining distress and symptomatology during the victims' rehabilitation process.

Traumatic events have the ability to influence and determine our thinking about the future and future adjustment in regard to mental and physical health as well as social adjustments. Many people use various coping strategies in order to obtain future orientated thinking and adjustment with the traumatic events. Traumatic events faced by adults are positively related to both future orientation and fear of future trauma that might arise (Holman & Silver, 2006).

Traumatic events often happen by accident and have the ability of changing the general life style profoundly which can result in long lasting psychological problems. It is difficult to establish a distinct definition for what a traumatic event is, as incidents can be perceived as traumatic for one individual but not for another. According to DSM-IV, a traumatic event is an experience that causes physical, emotional, and psychological distress or harm (American Psychiatric Association, 2000). In this case, a proper rehabilitation process is necessary. The rehabilitation processes that can be applied depend on the nature of the traumatic event and the social cultural context the victims come from and are part of. Regardless of the specific context, the rehabilitation process should have a high content validity in order to enable long term and effective results for the victims as traumatic events can have long lasting effects when not treated accordingly. Most of the victims who are affected by traumatic events show numerous emotional disturbances and damages resulting in drastic life changes. Among the most prevalent symptoms of traumatic events are behavioural and mood changes, such as sleeplessness, loss of appetite and emptiness of facial expressions. However, the underlying emotions that cause the superficial reactions that can be observed are not easily revealed. Therefore, understanding patients' emotions and using their emotional status in order to facilitate reliable coping strategies is essential. As the World Health Organization (WHO) points out a rehabilitation process can be defined as the provision of services consistent with the level of impairment, disability and handicap relative to the patient's personal preferences, needs and resources. Rehabilitation processes can be either represented institutionally (IBR) or community based (CBR).

3 Method

This research will be conducted with a sample from Germany and Sri Lanka. In total, 400 male and female university students ranging from 18 to 24 years of age will be represent the sample (200 from Sri Lanka and 200 from Germany). Identification of the dimensions of Emotional Intelligence will be measured by Trait Emotional Intelligence Questionnaire (TEI-Que) (Cooper & Petrides, 2010; Petrides & Furnham, 2006), while coping strategies are to be measured by with Brief Coping (Carver, 1997). Students resilience capacity which means the one's ability to adapt to stressful situations or crisis situations will be measured with by Resilience Scale for Children & Adolescents (Emotional Reactivity scale) (Prince-Embury, 2005). Effects of traumatic events faced by the students are to be identified with by the Essener Trauma- Inventor (Tagay, Duellmann, & Senf 2009). Data will be collected from the German student sample first using this questionnaire. Questionnaires will then be translated into Sinhala (National language in Sri Lanka) in preparation for data collection from the Sri Lankan students.

4 Discussion

Trauma is an issue, not only came afterwards natural disasters but also a commonly indicated issue even in our daily life. Especially, when youths are dealing with a new contextual orientation they have to come up with different types of traumas. Some of them have come across traumas in their life but not have been identified or treated. But the consequences and side effects of those traumas can be still visibly or hidden manner emerge among them. Trauma among university students not only impact upon academic performances but also towards the negative adjustment within their academic performances and personal life. Cross-cultural comparison of this research to be done in between Germany and Sri Lanka would be greater resources to identifying the cultural variations of different sources of trauma and the coping strategies based upon Emotional Intelligence and Resilience capacities used by the two different populations.

5 Affiliation

M.Sc. Asanka, Bulathwatta
Institution: University of Leipzig, Educational and Rehabilitation Psychology
Address: Neumarkt 9–19, 04109 Leipzig, Germany
E-mail: asankabulathwatta@gmail.com

6 References

Holman, E. A., & Silver, R. C. (2006). Future-oriented thinking and adjustment in a nationwide longitudinal study following the September 11th terrorist Attacks. *Motivation and Emotion, 29*(4), 389–407.

La Greca, A. M., Silverman, W. K., Vernberg, E. M., & Prinstein, M. J. (1996). Symptoms of postraumatic stress in children after Hurricane Andrew: A prospective study. *Journal of Consulting and Clinical Psychology, 64*(4), 712–723.

Mayer, J. D., Salovey, P., & Caruso, D. (2000). Models of Emotional Intelligence. In Robert J. Sternberg, *Handbook of Intelligence* (pp. 396–420). Cambridge: University Press.

Petrides, K. V., Cooper, A., & Furnham, A. (2006). *Ability and Trait Emotional Intelligence Questionaire.* London: Psychiatric Laboratory.

Prince-Embury, S. (2005). Resilience Scale for Children & Adolescents, Psychological Symptoms and Clinical Status in Adolescents. *Canadian Journal of School Psychology, 23*, 41–56.

Salovey, P. & Mayer, J. D. (1990). Emotional Intelligence. *Imagination, Cognition and Personality, 9,* 185–211.

Silver, R. H., Holman, E. A., McIntosh, D. N., Poulin, M., Gil-Rivas, V., & Pizarro, J. (2006). A nationwide longitudinal study of responses to the terrorist attacks of September 11, *The Journal of the American Medical Association, 288*(10), 1025–1041.

World Health Organization (WHO). (2012, May 7). *Disaster, Disability and Rehabilitation.* Department of Injuries and Violence Prevention, Geneva, Switzerland. Retrieved from http://www.who.int/violence_injury_prevention/other_injury/disaster_disability2.pdf.

Verbra Pfeiffer[1] & Sivakumar Sivasubramaniam[2]

[1] University of Stellenbosch, South-Africa

[2] University of the Western Cape, South-Africa

First Year Students: Using Expressive Writing to Cope with Trauma

Abstract. This study investigated the use of expressive writing to help students deal with any trauma they may be facing. For many students being a first year university student can become a frightening experience. They are not only faced with the new environment, but they also have to deal with new subjects. Besides having to cope with this new social environment they find themselves in, very often students in South Africa come from struggling family environments. This study has been designed against a backdrop of many students coming from very wobbly family structures, where they are faced with the sad reality of broken homes, estranged parents, single parents and parents not being there for them. It will be helpful to view them through interrelated ways by which students behave and react to classroom attendance and in the doing of tasks in class. A qualitative research method was used, which consisted of interviews, journal entries and autobiographical writing. This study was conducted over the duration of a semester at the Cape Peninsula University of Technology. This research was meant to offer support to our premise that students can use expressive writing to deal with any emotional trauma they may be facing. This study, not only investigated the issue of language proficiency, but also paved the way for students to write down their emotions. The findings suggest that the students have benefitted from the use of various activities that they undertook in this study via their expressive writing. Based on the feedback received especially from the second interview and their journal entries, we confirm that the students have gained confidence in their writing about themselves.

Keywords: expressive writing, journal entry, autobiographical writing, emotional trauma.

1 Introduction

We wish to discuss the educational and social concerns that underpin our study, which we hope can act as an awareness- raising exercise and a point of departure for our research. It has been our observation that not only do our university students find it difficult to pass their exams, but they also have to deal with the new social environment, which they have entered into.

We believe that this alerts us to the presence of a way of thinking in our educational settings, which views educational practices in terms of a rationalistic-technological stance.

As an antidote to this malaise, we wish to propose a concept of literacy which encourages democratic and liberatory change, which enhances the possibility to educate our students about the dialectical relationships between them and the world on the one hand, and language and change on the other (Freire & Macedo, 1987). In light of this language pedagogies and practices that target and signify students' experience and response assume immediacy and primacy. We believe that such pedagogies will be able to teach our students to assert their rights and responsibilities. We maintain that it will not only teach our students to read, understand and transform their own experiences but will also teach them to reconstruct their relationship with their society and that it will equip them better to process knowledge that is beyond their experience and to view their reading and writing as acts of empowerment (Freire & Macedo, 1987). The issues discussed so far, reflect our faith in the potential that reading and writing have for nurturing critical consciousness, especially when delivered through pedagogies of response. In order to promote learning through personal 'response' and experience, we particularly looked at journal entry and autobiographical writing. By using this form of writing, our intention was to awake in students their joys, fears, sorrows, abstractions, hopes and intuitions, to help them – students – become better writers and citizens.

By addressing linguistic, methodological and pedagogical issues and the corresponding values that accrue from them, our research attempted to appraise the use of interviews, journal entry and autobiographical writing as a means of promoting student-centered pedagogies and practices in writing. The rationale for our investigation was to use these literary methods, which we believe is fundamental to expressive writing.

We were curious to find out if the use of journal entry and autobiographical writing will:

1. motivate students to become better writers.
2. promote learning through 'response'.
3. help them emotionally.

At this point we wish to argue that based on the daily living experiences that students go through it may be able to provide provisional interpretations through writing which can bring about constructive educational and social change. The envisaged scheme of investigation used a qualitative research methodology.

2 Literature

This study aimed to demonstrate that writing is an essential skill needed to help in many aspects of 'oneself', whether emotionally or academically. The philosophical and educational foundations of expressive writing and its development as a pedagogical tool have been examined in this study through the technique of expressive thinking. Students needed to understand that expressive writing is NOT so-called 'creative writing' in which the writer essentially 'plays' without purpose or structure (Foulk & Hoover, 1996). It is not merely the act of thinking on paper about something which one probably does every day in the course of one's research, composition, and planning processes, it also deals with observations, analyses, and insights designed for a writer's personal use (Foulk & Hoover, 1996). We surmised that expressive writing is a manner of making connections between the 'known' and the 'new' on paper and it can also be defined as writing for the purpose of displaying knowledge or supporting self-expression (Graham & Harris, 1989; Russell, Baker, & Edwards, 1999). Foulk and Hoover (1996) define expressive writing as follows: expressive writing is writing in which the writer is her/his own audience. It needs to be evaluated by no one other than the writer. The expressive writing paradigm as understood by psychologists asks participants to write generally about their thoughts and emotions regarding traumatic life experiences, but researchers have used a variety of writing prompts, such as writing about life goals, one's best possible self, or an imagined traumatic event (Henry, Schlegel, Taley, Molix, & Bettencourt, 2010). In this sense, expressive writing has not been used to improve their writing, but rather their emotions. Researchers argue that on a personal level, the ability to express oneself in writing is important because writing about one's emotions can positively affect overall health (Gortner, Rude, & Pennebaker, 2006).

Ivanič (1998) identified three different but correlated selves that are socially constructed: autobiographical self, discoursal self and self as author. The "autobiographical self" "emphasizes writer's sense of roots" and "is itself socially constructed and constantly changing as a consequence of their [writers'] developing life-history" (Ivanič, 1998). "Discoursal self" is identified as "the impression – often multiple, sometimes contradictory- which they [writers] consciously or unconsciously convey of themselves in a particular text" (Ivanič, 1998). The third way of regarding writer's self in the writing act is the "self as author" which refers to the extent at which the writer perceives his/her self as an author, as well as it "concerns the writer's 'voice' in the sense of writer's positions, opinions, and beliefs" (Ivanič, 1998). Allowing learners the opportunity to explore their autobiographical self in writing, can make writing a meaningful and an empowering experience. In this sense, writing helps learners to connect their identities, understand their discoursal self, and thus develop their sense of authority in writing (Park, 2013b).

According to Pennebaker and Chung (2007) one of the basic functions of language and conversation is to communicate coherently and understandably. They mentioned that writing about an emotional experience in an organized way is healthier than in a chaotic way. They were relating writing to the emotional impact it has on the writer. In this study we hope to show how writing has affected our students' emotionally. There has been growing evidence from several labs which suggest that people are most likely to benefit if they can write a coherent story and that any technique that disrupts the telling of the story or the organisation of the story is undoubtedly detrimental. A writing exercise was conducted where they gathered people to write diary entries in first person, then write about the same event in the second person perspective and finally the people had to write that same event from a third person perspective. They noticed that these changes in writing perspectives were more of an emergent property of successful writing. In light of this, we believe that expressive writing brings about changes in people's social lives and has shown to increase working memory (Pennebaker & Chung, 2007). We also believe that expressive writing is relevant to work in autobiographical memory where you are forced to stop and look back at your life and evaluate issues and events that have shaped who you are, what you are doing and why. This study expected to demonstrate the above mentioned position as we conducted an autobiographical exercise with our students.

3 Method

3.1 Sample

As mentioned earlier, our methodology was predicated on a qualitative design. It should be noted from the outset that this is not an empirical design, and it is not our intention to compare and contrast students' writing. Rather the idea is to thematically and qualitatively account for the strategies used to establish their emotional stability. Our study, therefore, was designed to describe qualitatively as well as impressionistically the kinds of writing processes and strategies, and the outcomes arising out of them.

3.2 Measurement tools

Our data was collected from a literature-based programme, which we had implemented at our undergraduate wing of the faculty as the mainstay of our investigative study. In our study we used 14 students and our data collection was spread over a semester. Our study has identified one barrier, which has not been addressed so far and that is: neglect of expressive written English and a lack of confidence to write it. The need to address this barrier has prompted us to use journal entry and autobiographical writing, with a view to find possible solutions to students expressing them at the university. Some of the concerns that we liked to address in our study were:

1. Emphasize writing as a leeway for the exercise of mental energy and creativity.
2. Instil a love for writing in students by which it is hoped, it will encourage them to write for themselves both for their enjoyment and for their longer term benefit, as their range of writing expands.
3. Help students discover the personal utility of writing and the life-long joys and delights associated with writing (Sivasubramaniam, 2004).

Our research expected to demonstrate that the realization of expressing themselves in writing would help the learners minimize their debilitating anxiety, a feeling of deprivation which gets in the way of learning, and optimize facilitating anxiety, a feeling that making a real effort might make all the difference between success and failure, and thus help them do better than they might otherwise (Kleinman, 1977; Scovel, 1978, as cited in Sivasubramaniam, 2004).

Journal entry. When writing about themselves students wrote with greater fluency and satisfaction when their writing involved them personally, while they wrote with less facility when the writing was more objectified (Britton, 1980). In their journal entries students have been able to ask questions, admit confusion, make connections, and grow ideologically (Good & Whang, 1999 as cited in Cisero, 2006: 231). The entries of self-report (Bailey & Nunan, 1996) were to be a mixture of diary-like accounts of day-to-day experiences, in which students would reflect on themselves and what experiences they encountered during the day. Our intention with the journal writing endeavours were to provide a basis for free and expressive writing, which focused on developing fluency, the journals were however, never marked, but looked at regularly.

Autobiographical writing. This type of writing was intended to provoke personal response, critical appreciation and imaginative use of the language. The students were given samples of autobiographies, so that they had an idea as to the style of writing. This style of writing was appropriate because it helped the students to be creative and this method helped them to express themselves in writing.

Interviews. An interview was conducted at the beginning of our data collection and at the end of our data collection. It was administered to selected students who were participating in our study. They were selected from four different first year Communication classes. We chose a structured interview format for our undertaking but ascertained that there was adequate scope for using open-ended questions and open-ended responses.

4 Results

Our data analysed on autobiographical writing demonstrated that by using elaborative processing the students appeared to apply new information to their own lives in order to relate meanings to experience. We surmised that this method helped them translate new information into new formations while their elaborative processing helped them conceptualize their encounters with it. With the data we collected it became evident that our students have been able to foster the ability to construct an environment that was imaginative and assign a location to that environment in their real world through the use of writing. Their journal entries and autobiographical writing appear to have provided an emotional release and a digressing viewpoint signifying their continuing involvement with their daily living. The level of engagement appears to have been so gratifying that never once did they show any sign of tedium or demotivation.

We found that if expressive writing is to be banished for the sake of promoting cognitive and thought maturity, then it might result in a kind of writing that is lacking of human emotions and feelings. From a literary perspective, our students used their writing to question their own social identity, and therefore they attempted to develop new conceptual ways of thinking about themselves, their world and the 'others' in it (Barro, Jordan, & Roberts, 1998). The analysis of the students' journal entries and autobiographical writing indicated how through personal constructions, the students have attempted to relate the texts to their own emotions and relationships. By using entries of self-report to promote expressive thinking, our students have been able to discover the importance and inspiration to continue writing in a journal. This can be confirmed with reference to the response of the students in the second interview. We surmise that writing in journals can be completely open-ended and happen during a time when students can freely express themselves (Peyton & Read, 1990). While conducting the interviews, we were under the impression that students with a sense, wrote with some kind of freedom about what they were feeling and experiencing. Speaking in a journal helped students get their meanings across clearly. Data from the journal have served to legitimate the meaning constructions of the students. The data can serve to explain the effectiveness of the space provided by the journal for the students to democratize their personal experiences and in doing so find a basis for their intellectual, emotional and critical growth. Writing very personal things as they have done must have given them some kind of 'freedom' in writing. Autobiographical reflection (Galindo & Olguin, 1996; Pavlenko, 2003) appears to have helped them recapture personal experiences with Otherness and being othered themselves. Data from the interviews provided the required attitudinal flavour to the inquiry in that it brought to light the students' account of 'their lived through experience' of daily living. We believe that through the act of writing itself, ideas are explored, clarified, and reformulated and, as this process continues, new ideas suggest themselves and become assimilated into the developing pattern of thought (Zamel, 1983).

5 Discussion

We believed in encouraging our students to feel free to voice and address their fears, joy, hopes, doubts and initiatives which we believe would serve as the route of their expressive writing. Our choice of research design and methodology in our investigation was meant to capture the essence of the 'response' phenomenon in its fitting details and our aim was to provide a fuller explanation of it. Given that quantitative/reductive methodologies often failed to provide a fuller account of the phenomenon in focus, our study attempted to overcome that drawback by using a qualitative methodology of our informed choice. Our students' writing as seen recorded in our data illustrated what their attempts at input generation, meaning negotiation and motivational benefits coming as a result of their writing endeavours. Our intention was that the interviews, journal and autobiographical writing would become avenues for exploring meaning. We found that there was a kind of emotional release that came as a result of students' expressing themselves in their writing. Expressive writing encouraged them to learn through a response of dealing with topics which dealt with daily living, like fears, joy, hopes, doubts, initiations, intuitions which constitute the route of expressive writing. This is synonymous with the autobiographical writing that our students did. Though which they were able write about their emotions and relationships. Moreover, we believe that journal entries offered verifiable support as to how the students used their writing as a basis for thinking about different aspects of dealing with their situations. An aspect of the time limitation was the length of the study itself.

Our impression was that one semester was simply not enough time to gauge statistical improvements needless to say that we are not devotees of statistical truths and confirmations. First of all, the students reported that they loved writing in the journals. It increased their confidence, and many asked if they could keep their journals when we were done with them, to look back on it. We observed that the students used the journals to reflect on their lives. We got to know the students significantly better from what they wrote in their journals. It might be useful if first year students kept a journal for an entire year and not just for a semester. Doing this entire investigation has been a valuable learning experience for us. Our research methodology has made us more aware of the activities we chose to do with our students in classroom, which has caused us to question the measurability of expected results and also to appreciate the importance of follow through. We hope that tertiary students everywhere can look at this study and consider the immeasurable benefits of using journal entries and autobiographical writing.

6 Affiliations

Dr. Verbra Pfeiffer
Institution: University of Stellenbosch
Address: GG Cilliers Building, Ryneveld Street, 7602 Stellenbosch, South-Africa
E-mail: VFPFEIFFER@sun.ac.za

Prof. Dr. Sivakumar Sivasubramaniam
Institution: University of the Western Cape
Address: Robert Sobukwe Road, 7585 Bellville, South-Africa
E-mail: vfpfeiffer@sun.ac.za

7 References

Bailey, K. M., & Nunan, D. (1996). Introduction. In K. M., Bailey. & Nunan, D. (Eds.). *Voices from the Language Classroom* (pp. 1–10). Cambridge: Cambridge University Press. Pp. 1–10.

Barro, A., Jordan, S., & Roberts, C. (1998). Cultural Practice in Everyday Life: The Language Learner as Ethnographer. In M. Byram & M. Fleming (Eds). *Language Learning in Intercultural Perspective. Approaches through Drama and Ethnography*. Cambridge: CUP.

Britton, J. (1980). Shaping at the Point of Utterance. In A. Freedman & I. Pringle (Eds). *Reinventing the Rhetorical Tradition*. Las Vegas: Long and Silverman.

Cisero, C. A. (2006). Does Reflective Journal Writing Improve Course Performance? *College Teaching*, 54(2), 231–236.

Freire, P., & Macedo, D. (1987). *Literacy: Reading the Word and the World*. Connecticut: Bergin and Garvey.

Foulk, D., & Hoover, E. (1996). *Incorporating Expressive Writing into the Classroom*. (Technical Report Series no. 16). Minnesota: University of Minnesota.

Galindo, R., & Olguin, M. (1996). Reclaiming Bilingual Educators' Cultural Resources: An Autobiographical Approach. *Urban Education, 31*, 29–56.

Gortner, E., Rude, S., & Pennebaker, J. (2006). Benefits of Expressive Writing in Lowering Rumination and Depressive Symptoms. *Behaviour Therapy, 37*, 292–303.

Graham, S., & Harris, K. R. (1989). Improving learning disabled students' skills at composing essays: Self-instructional strategy training. Exceptional Children. *National Institutes of Health, 59*, 201–214.

Henry, E. A., Schlegel, R. J., Taley, A. E., Molix, L. A., & Bettencourt, B. A. (2010). The Feasibility and Effectiveness of Expressive Writing for Rural and Urban Breast Cancer Survivors. *Oncology Nursing Forum, 37*(6).

Ivanic, R. (1998). *Writing and identity: The discoursal construction of identity in academic writing.* Amsterdam: John Benjamins.

Park, G. (2013). Writing is a way of knowing: Writing and identity. *ELT Journal, 67*, 336–345.

Pavlenko, A. (2003). I never knew I was Bilingual: Reimagining Teacher Identities in TESOL. *Journal of Language Identity and Education, 2*(4), 251–268.

Pennebaker, J. W., & Chung, C. K. (2007). Expressive Writing: Connections to Physical and Mental Health. In H. S. Friedman (Ed.), *Oxford Handbook of Health Psychology* (pp. 417–437). New York: Oxford University Press.

Menard-Warwick, J. (2008). The Cultural and Intercultural Identities of Transnational English Teachers: Two Case Studies from the Americas. *Tesol Quaterly, 42*(4).

Peyton, J. K., & Reed, L. (Eds.) (1990). *Dialogue Journal Writing with Nonnative English Speakers: A Handbook for Teachers.* Alexandria, VA: Teachers of English to Speakers of Other Languages, Inc.

Pfeiffer, V. (2014). *An Investigation of L2 Expressive Writing in a Tertiary Institution in the Western Cape.* (Unpublished doctoral dissertation). University of the Western Cape, Cape Town, South Africa.

Russell, G., Baker, S., & Edwards, L. (1999). *Teaching Expressive Writing to Students with Learning Disabilities.* Retrieved from http://www.ldonline.org/article/6201/?theme=print.

Sivasubramaniam, S. (2004). *An Investigation of L2 Students' Reading and Writing in a Literature-Based Language Programme Growing Through Responding* (Unpublished doctoral dissertation). University of Nottingham, Nottingham, U.K.

Zamel, V. (1983). The composing processes of advanced ESL students: Six case studies. *Tesol Quarterly, 17*(2), 165–187.

Juliet Roudini & Evelin Witruk

University of Leipzig, Germany

Consequences of Trauma Experience in Iran and some Middle East Countries

Abstract. Trauma is a worldwide problem, with severe and wide range of consequences for individuals and society as a whole. Middle East countries have been exposed to political violence, wars, and natural disasters within the past few years. There were some natural disasters such as Bam earthquake in Iran which destroyed the whole city (Argue Bam) in 2003. Some of the researchers have done noteworthy studies to help victims to deal with psychological, physical problems after the traumatic events in Middle East countries. The aim of this research is to find psychological consequences of natural disasters in recent investigations in middle east countries. This study is a research review based upon some significant studies explaining positive and negative consequences of trauma affected on Iranian and Middle East people. We founded in the most researches that the most recognized demographic risk factors for Post-Traumatic Stress Disorder (PTSD) onset are children, females and young people. Mainly, and consistent with other PTSD studies, the greatness of the exposure to the event is the strongest predictor of the expansion of PTSD. However traumatic events come up with several negative outcomes, still there are some evidences from some researches that trauma has some optimistic outcomes such as post-traumatic growth.

Keywords: disaster, trauma, posttraumatic stress disorder, earthquake.

1 Introduction

Millions people around the world are victims of physical or sexual abuse, living in the terrifying atmosphere of natural disasters, domestic violence, car accidents and all can have distressing influence on the children and adults. Iran has been exposed to more than 34 out of 41 types of known disasters caused by natural hazards (Khankeh, 2011). In December 26th 2003 at 5:26 a major earthquake (6.6 on the Richter scale) devastated Bam in south Iran. This city is well-known for its 2500-year-old historical Site "Argue Bam". In seconds the whole city and its historical monuments were destroyed; more than 40,000 of the 100,000 residents were dead and 30,000 were injured (World Health Organization [WHO], 2004). Some of investigation showed that the amount of mental health difficulties and psychological distress between survivors was unexpected and there is a serious need to provide mental health care to disaster sufferers and to decrease negative health impacts of the earthquake (Montazeri, 2005). The research of H. R Khankeh revealed that the essential basic needs should be noticed to deliver comprehensive recovery facilities. In this regard, one of the basic needs is the requirement of continuous mental health care in the community (Khankeh, 2013).

2 Literature

Natural disasters do not only affect economically, but they also have physical injuries and further mental difficulties for the wounded people. Posttraumatic stress disorder (PTSD) and other related disorders are mentally side effects of disasters. Disaster mental preparedness is a significant reduction method to protect individuals from detrimental psychological effects arising from unexpected natural disasters. The experience of dealing with the tsunami emphasized the fact that disaster preparedness strategies meet the mental health and psychosocial needs of the community (Htay, 2006).

Within the past few decades, Middle East countries have witnessed political violence, wars, natural disasters, occupation, with insufficient resources available to support survivors dealing with the resulting psychological, physical and financial aftermath of those stressful events. Some of the investigations have examined the occurrence of PTSD in the community of the Middle East. Numerous children lost one or two family member, siblings or extended family members. They witnessed the most traumatic event in their life and were not able and prepared to deal with the situation. PTSD was testified in 47 % of children and 77 % of adults, which showed high vulnerability to PTSD after disasters in Iran (Yasamy, 2003).

3 Results of related studies

In Lebanon, Karam (2008) found 3.4 % lifetime prevalence rate of PTSD in a nationally representative sample (n = 2,857) of adult Lebanese civilians. In this research individuals were interviewed using the completely planned by WHO (2009) Composite International Diagnostic Interview 3.0. Lifetime occurrence of any Diagnostic and Statistical Manual of Mental Disorders, (DSM-IV) disorder was 25.8 % with Anxiety disorder (16.7 %) and mood disorder (12.6 %) were more mutual than impulse control (4.4 %) and substance abuse disorder (2.2 %). In Israel, 9.4 % of the adult sample (n = 512) were found to have a PTSD diagnosis. In the Gaza Strip, researchers reported that the lifetime prevalence of PTSD among adults (n = 585) was 17.8 % (De Jong, 2001). One of the significant studies was found regarding prevalence of DSM-IV psychiatric disorders among

7 to 10 year old Yemeni school children (Al-Yahri, 2008). Representative samples of Yemeni 7–10 year old (n = 1,210) were evaluated using a two phase design in a municipal area and another phase designed in a countryside area. In this research, psychopathology was considered using the Strengths and Difficulties Questionnaire for screening aims and the Development and Health Assessment to produce psychiatric diagnoses. The whole prevalence of DSM-IV disorders shows 15.7 % in Yemen. The result of this study indicated that anxiety disorders were the most common diagnostic grouping in Yemen (9.3 %), consequently behavioral disorders (7.1 %) and the last one is attention-deficit/ hyperactivity disorder (1.3 %).

Some important researches in adult and children survivors of the 2003 Bam earthquake described high prevalence rates of PTSD among adults with 81 %; (Hagh-Shenas, 2006), and somewhat lower prevalence amounts among children (36 % and 52 % in children over and under 15 years old (Parvaresh, 2009). Parvaresh study determined the PTSD in Bam-survived students who settled to Kerman four months after the Earthquake. In this research body injury and losing family were associated with PTSD symptoms in students older than 15 years old. Female gender was correlated with posttraumatic stress disorder and behavioral problems in students younger than 15. PTSD was distinguished in 36.3 % of the students older than 15 years and 51.6 % of the students younger than 15. In this group of children, behavioral difficulties were present in 31.3 %. Many children lost one or two parents, siblings, or extended family members. They witnessed the most stressful event in their life and were not prepared to deal with the condition Posttraumatic stress disorder (PTSD) was reported in 47 % of children and 77 % of adults, which showed high vulnerability to PTSD after disasters in Iran.

Another study that focused on PTSD and general symptoms of anxiety in 11–18 year olds, demonstrated 45.1 % of this group of adolescent had PTSD. Result of this research shows that about half of the survivors of Bam Earthquake who were exposed to the trauma of Bam earthquake had PTSD in months 7–9, without differences due to ender. Nevertheless, the severity of anxiety disorder signs was bigger in girls. Consequence of this research could be supportive to estimate the frequency of PTSD symptom clusters and the severity of anxiety symptoms in a group of adolescents in months 7–9 after the earthquake (Mousavi, 2006).

The last study considered the association between resilience and religious orientation (internal and external) with posttraumatic growth (PTG) in Iran community. Results of investigation of E. Witruk show that people with a strong belief in a just world attribute to the hazards as a result of human failure and this belief would be an important factor when dealing with natural disasters (Witruk, 2014). Even though investigators and researchers expansively have more emphasis on the undesirable effects of trauma, little consideration has been paid to the opportunity of optimistic

Influence of negative events. The concept of PTG was defined as "the experience of significant positive change arising from the struggle with a major life crisis" (Calhoun, 2000). Sample of this study was students ($n = 210$) who had experienced a minimum of one traumatic event within the last five years. Students completed the Traumatic Life Event Questionnaire (TLEQ), the Posttraumatic Growth Inventory - Iranian type (PTGI-I), and the Religious Orientation Scale (ROS). Result of this research indicates that some subscales of resiliency and religious orientation could predict posttraumatic growth in Iranian sample and marriage had an optimistic effect on feeling greater level of PTG, beside that social support has a noteworthy role in posttraumatic growth. According to the data of this research, women group reported more growth than men. Openness to experience was fundamentally significant feature for suitable growth of individuals facing a trauma. Optimistic individuals presented more flexibility in their coping strategies, consequently had a tendency to familiarize themselves to interesting situations.

4 Discussion

Natural disasters have not only economic effect, but also physical injury, and suffering from mental difficulties. PTSD, anxiety and other related disorders are mental side effects of disasters. Numerous researches on disaster mental consequences have been undertaken in several countries, however, the available investigates about disaster mental health preparedness are few in number, especially in the countries that they are at high risk of natural disaster like Iran and other Asian countries. Disaster mental preparedness is a significant reduction method to protect individuals from detrimental psychological effects arising from unexpected natural disasters. We have an absence of policy and planning for people with an economic difficulty, individuals with mental disorders and specific population like children, women and elderly people. In addition, standard training exercises for general and specific populations are required.

5 Affiliations

M.A. Juliet Roudini
Institution: University of Leipzig, Educational and Rehabilitation Psychology
Address: Neumarkt 9–19, 04109 Leipzig, Germany
E-mail: juliet.roudini@studserv.uni-leipzig.de

Prof. Dr. Evelin Witruk
Institution: University of Leipzig, Educational and Rehabilitation Psychology

Address: Neumarkt 9–19, 04109 Leipzig, Germany
E-mail: witruk@uni-leipzig.de

6 References

Al-Yahri, A. G. (2008). The prevalence of DSM-IV psychiatric disorders among 7- to 10-year-old Yemeni. *Social Psychiatry and Psychiatric Epidemiology, 43,* 224–230.

Bleich, A. G. (2003). Exposure to terrorism, stress-related mental health symptoms, and coping behaviors among a nationally representative sample in Israel. *Journal of the American Medical Association, 290,* 612–620.

Calhoun L. G., C. A. (2000). A correlational test of the relationship between posttraumatic growth, religion, and cognitive processing. *Journal of Traumatic Stress, 13,* 521–527.

Calhoun LG, et al. (2000). *Behavioral Emergencies for the Emergency Physician.* Edingburg: Cambridge University press.

De Jong, J. T. (2001). Lifetime events and posttraumatic stress disorder in 4 post-conflict settings. *Journal of the American Medical Association, 286,* 555–562.

Hagh-Shenas, H. G. (2006). Post-traumatic stress disorder among survivors of Bam earthquake 40 days after the event. *Revue de Santé de la Méditerranée Orientale, 12,* 118–25.

Htay, H. (2006). Mental health and psychosocial aspects of disaster preparedness in Myanmar. *International Review of Psychiatry, 18(6),* 579–585.

Karam G, Z. N. (2008). Lifetime Prevalence of Mental Disorders in Lebanon: First Onset, Treatment, and Exposure to War. *PLoS Medicine, 5(4),* 0579-0586.

Khankeh, H. K.-Z. (2011). Disaster Health-Related Challenges and Requirements: A Grounded Theory Study in Iran. *Prehospital and Disaster Medicine, 26,* 151–158.

Khankeh, H. N.-Y. (2013). Life Recovery After Disasters: A Qualitative Study in the Iranian Context. *Prehospital and Disaster Medicine, 28(06),* 573–579.

Montazeri et al. (2005). Psychological distress among Bam earthquake survivors in Iran: a populationbased. *BioMed Central Public Health,* 1–6.

Parvaresh, B. (2009). Post-traumatic stress disorder in Bam-survived students who immigrated to Kerman, four months after the earthquake. *Archives of Iranian Medicine, 12,* 244–249.

S Mousavi, J. M.-G. (2006). Post Traumatic stress Disorder and General Symptoms of Anxiety in Adolescent Survivors of Bam Earthquake. *Iranian Journal of Psychiatry, 5(4)* 76–80.

World Health Organization (WHO). (2004). WHO joins international effort to help Bam earthquake survivors. *Bulletin of the World Health Organization, 82(2)*, 156–157.

Witruk, E. (2014). Dealing With Earthquake Disaster on Java 2006: A Comparison of Affected and Non-affected People. *Indonesian Psychological Journal, 29(3)*, 121–135.

Yasamy, B.-Y. (2003). *Psychosocial Support Interventions on Survivors of Qazvin Earthquake.* Tehran: Iran Red Crescent Society, UNICEF.

Hamidreza Khankeh[1,2], Amin Saberinia[3],
Davoud Khorasani-Zavareh[2,4], Ali Ardalan[5,6],
Maryam Nakhaei[7], & Maryam Ranjbar[1,8]

[1] University of Social Welfare and Rehabilitation Sciences, Iran

[2] Karolinska Institute, Sweden

[3] Kerman University of Medical Sciences, Iran

[4] Urmia University of Medical Sciences, Iran

[5] Tehran University of Medical Sciences, Iran

[6] Harvard University, USA

[7] Birjand University of Medical Sciences, Iran

[8] Institute of Humanities and Social Studies, Iran

Emergency and Disaster Health Provision in Iran: Challenges and Achievements

Abstract. Disasters are a significant threat to human life worldwide and can have a variety of destructive consequences for societies. Iran is one of the most vulnerable countries to different types of disasters such as earthquakes, floods, and droughts. Despite many achievements, concern raised regarding the effectiveness of disaster health management in Iran. This narrative review study aims to explore health disaster challenges and achievements in Iran through reviewing related studies published during the past ten years (1st June 2006 and 25th November 2014) and selected purposefully, regarding the challenges and achievements in emergency and disaster health provision in Iran. Manifest content analysis used to explore the challenges and achievements that the health system has experienced during recent Iranian emergencies and disasters. These include insufficient coordination in the provision of health services in emergency and disaster; lack of comprehensive National Health Emergency Management Plan (NHEMP); lack of a comprehensive recovery plan; and lack of a system approach to providing trauma management. Moreover, achievements covered introduction of a single national emergency number; improvement of the organizational level of Emergency Medical Services (EMS); public education; developing the Health Emergency Working Group responsible for all aspects of health disaster risk management in Iran. Despite many improvements in health provision in the field of emergency and disaster, there are still challenges considerable for both health policy makers and researchers. Duplication of efforts and wasting resources and time due to insufficient coordination and collaboration between the organizations involved needs further investigation.

Keywords: health, emergency, disaster, challenge, achievements.

1 Introduction

Devastation is the result of a disaster or catastrophic event, and consists of a combination of hazards, vulnerability, and insufficient capacity (Cox Jr, 2012). When a disaster happens, serious disruption occurs in a community or society. This causes widespread human, material, economical, and environmental loss. In this situation, the affected community or society finds that the challenge exceeds their ability to cope using only their own resources (Pollak, Born, Kamal, & Adashi, 2012).

The increased number of natural and manmade disasters all around the world lead to great suffering for human beings, and huge damage to properties and the environment. Large-scale disasters in particular have had great impact on public health provision and demonstrated that in the absence of a system approach for management, society will be highly vulnerable (Boin & Hart, 2010). Thirty years ago, social scientists developed a disaster risk reduction approach, the use of which has largely resulted in decreasing the disaster-related health consequences and vulnerabilities (Perry, 2007).

Globally more than 90 % of the affected populations and 50 % of both the death toll and economic losses accounted for by natural disasters in Asia. Iran is exposed to a wide range of natural and manmade disasters. "According to EM-DAT, 181 disasters were recorded in Iran from 1900–2007, which caused 155,811 deaths and 68,217 injuries, and affected 44,037,516 people." Earthquakes, droughts, and floods are the most serious events in Iran in terms of mortality, economic loss, and affected population. Iran is a country which has a very long history of dealing with the challenges of being exposed to more than 34 out of 41 types of the known natural disasters (Khankeh, Khorasani-Zavareh, Johanson, Mohammadi, Ahmadi, Mohammadi, 2011). In addition to disasters, Iran has one of the highest mortality rates from road traffic injuries (RTIs) among middle-income countries (Khorasani-Zavareh, Mohammadi, Khankeh, Laflamm, Bikmorad, & Haglund, 2009).

Related studies indicate that, the Islamic Republic of Iran requires the development of essential infrastructures to respond effectively to emergencies and disasters with appropriate preparedness and adequate health provision. After the Bam earthquake in 2003, the existing disaster-related laws revised and new regulations established but these legal supports and the related activities still need to be evaluated (Amini-Hosseini & Hosseinioon, 2012). Because Iran is one of the most disaster-prone countries in the world, the health and medical needs of the population is a major issue. Due to the lack of research in this important field, this study has been designed to explore the disaster health challenges and achievements.

2 Literature

This is a study, which is going to review the issues related to emergency and disaster health in Iran, focusing on challenges and achievements, no theory or literature will be mentioned in advance.

3 Method

3.1 Design

This is a narrative review study, which reviewed ten recently published studies regarding disaster health-related challenges and achievements in Iran.

3.2 Protocol

These studies published between 1 June 2006 and 25 November 2014; Both bibliographic and citation databases have been purposefully searched for articles studied the disaster health-related issues in Iran. Then the papers were reviewed for study eligibility, related to the research question "*what are the health disaster challenges and achievements in Iran?*"

3.3 Analysis

In this study, a qualitative manifest content analysis employed to explore the health disaster challenges and achievements; the analysis outcomes were the main health-related issues in disaster, while describing the current situation of disaster health provision in Iran. Health-related issues were then derived from the inductive process (Graneheim & Lundman, 2004).

4 Results

The search strategy and application of the inclusion criteria yielded 10 relevant health articles published within the mentioned period (Table 1). Reviewing these papers suggested the main challenges regarding health disaster management as follows: The lack of planning; inadequate organizational management of resources; insufficient coordination in the provision of health services in emergency and disaster; the lack of a single emergency number in emergencies (different emergency phone numbers), which probably results in duplicated efforts and wasting of resources and time; organizational conflicts; lack of a comprehensive NEMP[1]; lack of a system approach for providing trauma care; unclear national policies and poor organization of emergency management; inappropriate structure and unsupportive environment in the field of emergency management; lack of provision of comprehensive recovery services to meet the important basic needs of affected people after a disaster, including the need for physical and social rehabilitation and livelihood health; the need for continuity of mental health care; the need for family re-unification services; absence of a structured disaster plan; absence of standardized medical teams; shortage of resources; and finally insufficient policies to improve the quality of pre-hospital medical care and improve interaction between pre-hospital and hospital.

Table 1. Main Findings of relevant papers about Health in Disaster and Emergencies in Iran

Author	Title	Main Findings
Khankeh, H. R. et al. (2011)	Disaster Health-Related Challenges and Requirements: A Grounded Theory study in Iran	The main barriers in disaster health-related challenges, which hindered adequate disaster healthcare service delivery during the Iranian recent disasters.
Ardalan, A. et al. (2009)	Disaster Health Management: Iran's Progress and Challenges	The health system needs strengthening for intra and inter sectoral collaboration & coordination, information management system and community-based initiatives; and to focus on disaster risk reduction and enhancing response capacity.
Khorasani-Zavareh, D. et al. (2009)	The requirements and challenges in preventing of RTI in Iran. A qualitative study	Challenges in terms of RTI prevention identified. Suggestions made to improve, supervise and coordinate preventive activities.
Khankeh, H. R. et al. (2013)	Life Recovery After Disasters: A Qualitative Study in the Iranian Context	Main issues of a disaster recovery/rehabilitation, the comprehensive recovery services and important basic needs been considered. Social activation reintegrates affected people into the community.

---

1   National Emergency Management Plan.

Author	Title	Main Findings
Haghparast-Bidgoli, H. et al. (2013)	Exploring the provision of hospital trauma care for RTI victims in Iran: a qualitative approach	Major obstacles identified and building a national trauma system, using available professional resources, and implementing low cost and evidence-based improvements suggested as resolutions and developing strategies.
Djalali, A. et al. (2011)	Facilitators and obstacles in pre-hospital medical response to earthquakes: a qualitative study	The main barriers identified. The result showed that implementing a comprehensive plan would not only save lives but decrease suffering and enable an effective praxis of the available resources at pre-hospital and hospital levels.
Froutan, R. et al. (2014)	Pre-hospital burn mission as a unique experience: A qualitative study	Different factors affect the quality of pre-hospital clinical services for burns. The authorities should consider the physical and psychological health of staff and make policies to improve the quality of care.
Khankeh, H. R. et al. (2012)	Why the Prominent Improvement in pre-hospital Medical Response in Iran Couldn't Decrease the Number of Death -Related RTIs	Environmental factors showed as crucial in reducing road traffic injuries. Changing stakeholders' attitudes to develop road user safety, Safety education programs of modifying the behavior and improving the interaction within the current pre-hospital trauma care system should be emphasized as solutions.
Alipour, F. et al. (2014)	Challenges for Resuming Normal Life After Earthquake: A Qualitative Study on Rural Areas of Iran	Social uncertainty and confusion in the process of returning to normal life can greatly interrupt the normal development. Understanding the challenges of life recovery will help policy makers consider social rehabilitation as a key factor in facilitating the return to normal life after earthquakes.
Khorasani, D. et al. (2009)	Post-crash management of road traffic injured victims in Iran. Stakeholders' views on current barriers and potential facilitators	Barriers found as involvement of laypeople, lack of coordination, inadequate pre-hospital services, and shortcomings in infrastructure. Suggestions made as a public education campaign in first aid, the role of the emergency services, cooperation of the public at the crash site, target-group training for professional drivers, police officers and volunteers.

The main achievements explored as the National campaign for RTIs since 2005; establishment of one single national emergency number to integrate all dispatches; establishing an Emergency Management Center to improve the organizational level of EMS; and public education regarding family preparedness against disasters.

According to the legislations, the National Disaster Management Organization (NDMO), endorsed by the Parliament of Iran, there are now 14 Working Groups responsible for inter and intra-sectoral coordination of disaster risk management. The Ministry of Health and Medical Education (MOHME) is the leading agency and the location of the Health Emergency Working Group and its secretariat office, which consists of six committees: medical services, public health, logistics, education, research and rehabilitation. The corresponding departments of MOHME are in charge of these various committees. The Emergency Management Center (EMC) established in 2006 at the MOHME to be in charge of organizing, management, and coordination of emergency and disaster healthcare. Now all the Universities for Medical Education and Health Services (provincial and sub-provincial MOHME) have their own functional EMCs consist of the following entities: Emergency Medical Services and Operations Center, Health in Emergency and Disaster National and a National Passive Defense Working Group. MOHME also established an Emergency Operations Center (EOC) in 2006, which is an adequately staffed and equipped facility to manage emergencies.

Some national tools developed to assess the health risks such as the Hazard and Capacity Tool, Hospital Safety Index and Hospital Preparedness Tool. Meanwhile, a national guideline for Hospital Disaster Planning (HDP) developed and established in some hospitals as a pilot project. A National Emergency Operation Plan developed based on National Disaster Scenarios in public health, pre-hospital and hospital care. Finally, MOHME established a PhD Program of Health in Emergency and Disaster to develop knowledge, resolve problems and reduce health risks of disaster in Iran.

5 Discussion

This narrative study reviewed the challenges and achievements experienced in the health system of Iran during recent emergencies and disasters using a qualitative approach. Important challenges found as insufficient coordination and collaboration in the provision of health services in the event of emergency and disaster, which led to duplication of efforts, wasting resources and time. Also noted were the lack of a Comprehensive NEMP, the absence of a structured disaster plan and standardized medical teams, the lack of a system approach in providing trauma care and unclear national policies. Important achievements in the field of health in emergency and disaster include: the introduction of the single, national emergency number in an integrated dispatch system; improving the organizational level of EMS by establishing EMC; public education; and developing Health Emergency Working Group as the responsible authority for all aspects of health disaster risk management in Iran.

In terms of disaster health care services, Khankeh et al. (2011) indicated that the most significant barriers to adequate health care services during the past decade in Iran were the lack of planning, inadequate organizational management of resources, insufficient coordination in providing health services during the disaster, and the manner of participation of international relief efforts (Khankeh et al., 2011). Studies in other developing countries have suggested that inadequate resources is the major problem in providing disaster health services. By contrast, the main problem in Iran is the lack of coordination and preparedness planning. However, the establishment of the Health Emergency Working Group and its secretariat in the MOHME could help to improve the coordination between health-related organizations to some degree. The national preparedness planning such as the Emergency Operational Plan (EOP) and HDP have been recently developed. The coordination problem has also been mentioned in the other countries. For instance Zoraster (2006) suggests that there are wasted efforts, resources and funds as the result of these recurrent problems when responding to disasters. According to his study, coordination problems were very apparent in the response to the earthquake and tsunami in south East Asia in December 2004, one of the major natural disasters of recent times (Zoraster, 2006).

The insufficient intra and inter-sectoral collaboration and coordination; the lack of a comprehensive information management system; and insufficient community-based initiatives, also explored as challenges and emphasized by Ardalan et al. (2009). They recommended a focus on disaster risk reduction while enhancing the response capacity. The research-based investment would lead to quality decision-making in disaster health management according to the present review.

Regarding the important challenges, this review revealed some national interventions in order to improve the coordination and collaboration and to increase the health system capacity and preparedness. The Emergency Management Center (EMC) established in the MOHME for organizing, management, and coordination of health in emergency and disaster and the Emergency Operations Center, along with the creation of The Health Emergency Working Group and National Passive Defense Working Group are other important achievements. The Iranian health system has achieved considerable organizational, academic and administrative success in terms of emergency management and risk reduction in the health sector, according to the results of this study, but there is still a great room for improvement.

Khorasani-Zavareh et al. (2009) and Haghparast-Bidgoli et al. (2013) argue that "the lack of a system approach to road-user safety" and "lack of a system approach in providing trauma care at emergency system" are two of the most important challenges regarding post-crash management in Road Traffic Injuries, which is a major issue in Iran (Khorasani-Zavareh et al., 2009; Haghparast-Bidgoli et al.,

2011). They suggested public education improvements, more effective legislation, more rigorous law enforcement, improved engineering in road infrastructures, and an integrated organization to superintend and coordinate the preventive activities (Khorasani-Zavareh et al., 2009; Khorasani-Zavareh et al., 2009).

The Iranian health system has had some achievements regarding these issues; for instance following the establishment of the National Campaign for RTIs since 2005 mortality has declined for 10 % each year respectively. Moreover, in 2014, the National Road Safety Committee, MOHME, approved the consolidation of the various dispatch systems of the different involved organizations into a single national emergency number (115) in order to improve a system approach, inter and intra organizational coordination and collaboration and to decrease the waste of resources and time and the duplication of the efforts.

Despite the fact that Iran has one of the highest mortality rates from RTIs in middle-income countries, it is believed that some national interventions in the field of Emergency Medical Services can change the situation. Studies on post-crash events in Iran, suggest that although about 60 % of deaths have previously occurred at the crash scene or during the transportation to the hospital, the pattern has changed in recent years. Nowadays around 40 % of the road traffic victims die in the pre-hospital phase. This pattern may have changed because of higher capabilities in the pre-hospital medical response and a dramatic improvement in staff training leading to a boost in skills. This has occurred in combination with an increase in the number of ambulances and dispatch sites and an upgrading of their equipment. These efforts may successfully decrease the mortality and morbidity of RTIs by a noticeable amount. All this can be attributed to some improvements at the organizational level of the EMS along with public education (Khankeh et al., 2012; Froutan et al., 2014).

This purposeful narrative review has revealed the main problems of the Iranian health system when facing the managerial issues of emergencies and disasters. Insufficient coordination when providing health services in the case of emergency and disaster, insufficient collaboration and lack of planning in the field of emergency services may tend to duplication of efforts, wasting resources and time. These important problems perhaps resulted from the lack of a Comprehensive NEMP. The involvement of different organizations with different emergency phone numbers and dispatch systems has been considered another barrier diminishing the efficiency of the emergency services. The health system has provided some national interventions to overcome these challenges and improve the coordination and collaboration as follows:

- Developing a single, national emergency number and the integrated dispatch system
- Improving the organizational level of EMS by establishing EMC and National Emergency Operation Center (EOC)
- Providing public education
- Developing an emergency and disaster health national working group responsible for all aspects of health disaster risk management in Iran

These efforts could decrease the mortality and morbidity of RTIs and natural disasters. Despite these national achievements, there is still a lot to be improved. The health services of disaster-affected countries should consider appropriate communication strategies with the relevant agencies to obtain international humanitarian assistance in emergencies and disasters.

The findings of the current study show that it is of the utmost importance that efforts and actions focused on the disaster risk reduction should be coordinated even more in order to restrict the impacts of disasters and to achieve recovery as soon as possible. The health system should be strengthened in intra and inter-sector collaboration and coordination, information management systems and community-based initiatives for disaster preparedness. It is also essential to focus on disaster risk reduction while enhancing the response capacity. Investing in research would lead to quality decision-making in disaster health management. Therefore, the key factors to develop an effective pre-hospital trauma care system, are improving the interaction within the current pre-hospital trauma care system and building a common understanding about the role of Emergency Medical Services. Moreover, the authorities and health system administrators should consider the physical and psychological health of their staff and establish policies to improve the quality of pre-hospital medical care. The attitude among stakeholders in relation to road safety activities must change. The focus of all activities should be on road users' safety. It is necessary to build a national trauma system, use available professional resources, and implement low cost and evidence-based improvements such as establishing trauma teams and trauma training for ED staff on a regular base in order to improve the trauma care at the hospitals. Finally, providing comprehensive recovery services and attending to the important basic needs of affected people should be considered, including the need for physical rehabilitation, social rehabilitation, and livelihood health; not to mention the need for continuity of mental health care and for family reunification services.

6 Corresponding author

Dr. Hamid Reza Khankeh
Institution: University of Social Welfare and Rehabilitation Sciences
Address: Koodakyar Ave., Daneshjoo Blvd., Evin St., Tehran, Iran
E-mail: hamid.khankeh@ki.se

7 References

Alipour, F., Khankeh, H. R., Fekrazad, H., Kamali, M., Rafiey, H., Sarrami Foroush-ani P. Ahmadi, S. (2014). Challenges for Resuming Normal Life after Earthquake: A Qualitative Study on Rural Areas of Iran. *PLoS Currents*, 17; 6.

Amini-Hosseini, K., & Hosseinioon S. (2012). Evaluation of Recent Developments in Laws and Regulations for Earthquake Risk Mitigation and Management in Iran. *Risk, Hazards & Crisis in Public Policy*, 3, 1–20.

Ardalan, A., Masoomi, G. R., Goya, M. M., Ghaffari, M., Miadfar, J., Sarvar, M. R., Soroush, M., Aghazadeh, M., (2009). Disaster health management: Iran's progress and challenges. *Iranian Journal of Public Health, 38*, S1, 93–7.

Boin, A., & Hart, P. T. (2010). Organising for Effective Emergency Management: Lessons from Research1. *Australian Journal of Public Administration*, 69, 357–371.

Cox Jr, L. A. T. (2012). Community resilience and decision theory challenges for catastrophic events. *Risk analysis, 32*, 1919–1934.

Djalali, A., Khankeh, H., Oehlén, G., Castrén, M., & Kurland, L. (2011). Facilitators and obstacles in pre-hospital medical response to earthquakes: a qualitative study. *Scandinavian Journal of Trauma, Resuscitation and Emergency Medicine*.

Froutan, R., Khankeh, H. R., Fallahi, M., Ahmadi, F., & Norouzi, K. (2014). Prehospital burn mission as a unique experience: A qualitative study. *Burns, 40*, 8, 1805–12.

Graneheim, U. H. & Lundman, B. (2004). Qualitative content analysis in nursing research: concepts, procedures and measures to achieve trustworthiness. *Nurse education today*, 24, 105–112.

Haghparast-Bidgoli, H., Khankeh, H. R. Johansson, E., Yarmohammadian, M. H., & Hasselberg, M. (2013). Exploring the provision of hospital trauma care for road traffic injury victims in Iran: a qualitative approach. *Journal of Injury and Violence Research, 5(1)*, 28–37.

Khankeh, H. R., Khorasani-Zavareh, D., & Masoumi, G. R. (2012). Why the Prominent Improvement in Pre-hospital Medical Response in Iran Couldn't Decrease the Number of Death Related Road Traffic Injuries. *Journal of Trauma and Treatment, 1.4.*

Khankeh, H. R., Khorasani-Zavareh, D., Johanson, E., Mohammadi, R., Ahmadi, F., & Mohammadi, R. (2011). Disaster Health-Related Challenges and Requirements: A Grounded Theory Study in Iran. *Prehospital and Disaster Medicine, 26(3)*, 151–8.

Khankeh, H. R., Nakhaei, M., Masoumi, G., Hosseini M., Parsa-Yekta Z., Kurland L., & Castren, M. (2013). Life Recovery after Disasters: A Qualitative Study in the Iranian Context. *Prehospital and Disaster Medicine, 28(6)*, 573–9.

Khorasani-Zavareh, D., Khankeh, H. R., Mohammadi, R., Laflamme, L., Bikmoradi, A., & Haglund, B. JA. (2009). Post-crash management of road traffic injury victims in Iran. Stakeholders' views on current barriers and potential facilitators. *BioMed Central Emergency Medicine*, 9: 8.

Khorasani-Zavareh D., Mohammadi R., Khankeh H. R., Laflamme L., Bikmoradi A., & Haglund B. J. (2009). The requirements and challenges in preventing of road traffic injury in Iran. A qualitative study. *BioMed Central Public Health, 23*, 9, 486.

Perry, M. (2007). Natural disaster management planning: A study of logistics managers responding to the tsunami. *International Journal of Physical Distribution & Logistics Management, 37*, 409–433.

Pollak, A. N., Born, C. T., Kamal R. N., & Adashi, E. Y. (2012). Updates on disaster preparedness and progress in disaster relief. *Journal of the American Academy of Orthopaedic Surgeons*, 20 Suppl 1, 54–8.

Zoraster, R. M. (2006). Barriers to disaster coordination: Health sector coordination in Banda Aceh following the South Asia Tsunami. *Prehospital and Disaster Medicine, 21*,

Nadia Hanum & Konrad Reschke

University of Leipzig, Germany

Indonesian Driver's Behavior: Post-Traumatic Stress Disorder (PTSD) after Accident caused by Bad Driving Practice

Abstract. In the year 2014 a study of Indonesian driver's behavior took place. The difficult driving situation and high number of traffic accidents in Indonesia were described. There is no developed Traffic Psychology in Indonesia. Also, there is a lack of methods to analyze the driving behavior. That's why the Manchester Driving Behavior Questionnaire was translated and used to carry out a Pilot study to find out some results of the quality and problems of Indonesian drivers. The Manchester Driving Behavior Questionnaire was answered by 299 drivers of three cities (regions) in Indonesia. The driver characteristics were assessed by an additional form. The first time driver behavior of Indonesian drivers (car, bus, truck, motorcycle and bicycle) was described by the scales: Aggressive Violation; Ordinary Violation; Errors; Lapses. One result was accidents caused by bad driving behavior can lead to Post-Traumatic Stress Disorder. Other results will be discussed shortly.

Keywords: traffic and transportation psychology, driving behavior, trauma.

1 Introduction

Indonesia covers a total area of 9.8 million square kilometers (km²). As an archipelago, it comprises a sea area of 7.9 million km (including an exclusive economic zone), or 81 % of the total area, and a land area of about 1.9 million km2. It is also a country with many volcanoes and rivers. The total population of Indonesia, according to the 2004 Population Census is 227 million (Ministry of Transportation of Indonesia, 2006).

Indonesian Ministry of Transportation and Ministry of Environment at 4th EST Forum in Asia (2009) described the current issue on the land transport sector. The issues are such as the high growth of urban-rural population and vehicle, traffic congestion, public transportation and air population. The number of urban-rural population and vehicle in ten years increased rapidly. By increasing population and number of vehicles in urban areas can cause problems on the highway as traffic congestion. Other factors that caused traffic congestion are such as Physical bottlenecks, capacity reduction at intersection, loading and unloading of bus passenger on the road, U-turns, railroad crossing, and bad driving practice (Indonesian Ministry of Transportation & Ministry of Environment at 4th EST Forum in Asia, 2009).

2 Literature

As many other countries may recognize that there are three main causes of traffic accidents namely, human factors, vehicle factors, and road and environmental factors, Indonesia also has there. It is, however, true that non-human factors in Indonesia have a greater percentage as compared to other countries, figures, and implicitly indicate human errors too, such as ignorance of human to vehicle and road maintenance (Soehodho, 2009). Accidents caused by human error or due to poor driving behavior on the road was a lot going on, but unfortunately these cases get a little portion in a study that the lack of reference that can be the basis of this study. Examine and analyze the phenomena of human error or human behavior on the road is one of the objectives of transport psychology.

Several studies have shown that traffic accidents are a common cause of post-traumatic stress disorder (PTSD). Ursano et al. (1999) and Bryant and colleagues (2004) found a prevalence of 25 % PTSD three months and 18 % six months after the traffic accident. PTSD seems to be an important psychological consequence of accidents with motorized vehicles. Most studies involve populations of patients selected according to the kind of injury caused by the accident, e.g. an orthopedic trauma (Starr et al., 2004), a spinal cord trauma (Nielsen, 2003) or a brain trauma (Harvey, 2000 in European Transport Safety Council, 2007).

European Transport Safety Council (2007) also reported that a previous trauma does not seem to be a risk factor (Ursano et al., 1999), although a previous episode of PTSD does. Richmond and Kauder (2000) identified four variables that were important in the prediction of psychological distress after a serious injury, namely increased levels of psychological distress during hospitalization, a positive screen for drugs and alcohol at the time of the injury, young age and the lack of anticipation of possible problems that can occur with when resuming normal activities. Zatzick (2002) examined 101 surgical inpatients and found that 73 % perceived a high level of psychological stress and/or were positive for intoxication with stimulants. One, four and twelve months after the injury, 30 to 40 % of the patients reported symptoms of PTSD. Severe symptoms in the beginning were the strongest predictor of continuing PTSD-symptoms during the following year. This suggests that one can assess predictors of PTSD from the moment of hospitalization and thus allow early assessment for referral into psychiatric care.

3 Method

3.1 Sample

In this study, the researcher takes three sub-samples. The sample is divided into three cities (Jakarta, Semarang and Yogyakarta) in the island of Java in Indonesia. Each city has its own characteristics that will be representing the characteristics of the population of Indonesia in general. The Manchester Driving Behavior Questionnaire had 299 respondents (drivers in each city).

3.2 Mesurement tools

There is cross-cultural questionnaire for a comprehensive model of everyday driving behaviors (Reason, Manstead, Stradling, Baxter, & Campbell, 1990). The Driver Behavior Questionnaire (DBQ) (Reason, Manstead, Stradling, Baxter, & Campbell, 1990) is one of the most widely used instruments for measuring driving style. The DBQ is based on a theoretical taxonomy of aberrant behaviors and the main idea in the DBQ being the distinction between errors and violations (Reason et al., 1990).

4 Results

4.1 Inference statistical analysis of differences in DBQ by age, sex and vehicle

We used one way variance analysis to detect differences in driving behavior between age, sex and vehicle use. The result of ANOVA was that male and female have no significant differences in DBQ's total value. The difference is only showing a direction, because the significance value of sex is only .08. In the age group, we found high significant differences in the total value of DBQ with .00. From the results the younger drivers show a higher negative degree in the dimension of driving behavior. It means that younger drivers show more aggressive violations, ordinary violations, errors and lapses. In the third step of this analysis, we looked for the influence of the used type of vehicle in relation to the driving behavior. There were no significant differences between the type of vehicle and the DBQ dimensions of driving behavior with .24.

4.2 Results of the cluster analysis

We computed cluster analysis by the program. The number of the clusters should be 3. We used the cluster analysis in SPSS 21 for windows. By the help of cluster analysis, the respondents could be classified into three clusters. Namely:

1. The cluster 1 ($n = 98$) consists of drivers who has lower scores in DBQ
2. The cluster 2 ($n = 63$) consists of drivers who has middle scores in DBQ
3. The cluster 3 ($n = 135$) consists of drivers who has higher scores in DBQ

The most persons from our sample belong to the Cluster 3. It means, the members of cluster 3 shows bad driving behavior, according to all dimensions of the DBQ, we have to take into account that the data are self-report data.

5 Discussion

From the results, it can be seen that the Indonesian driver's behavior are included in the category of bad drivers. Bad driving behaviors can increase the risk of accidents caused by human error. The effects of the accident are not only the damage to the road infrastructure and financial losses but also can provide trauma to the victim.

According to DSM-IV-TR (American Psychiatric Association [APA], 2000) specifically defines a trauma as direct personal experience of an event that involves actual or threatened death or serious injury, or other threat to one's physical integrity; or witnessing an event that involves death, injury, or a threat to the physical integrity of another person; or learning about unexpected or violent death, serious harm, or threat of death or injury experienced by a family member or other close associate (Criterion A1). The person's response to the event must involve intense fear, helplessness, or horror (or in children, the response must involve disorganized or agitated behavior) (Criterion A2, p. 463). DSM-IV-TR also provides a list of potentially traumatic events, including combat, sexual and physical assault, robbery, being kidnapped, being taken hostage, terrorist attacks, torture, disasters, severe automobile accidents, and life-threatening illnesses, as well as witness death or serious injury by violent assault, accidents, war, or disaster.

The effect of trauma from an accident victim can also be seen on the severity of the injuries suffered by accident victims. Injury due to an accident can be divided into two, namely mild injury and severe injury. Physically, the victims with severe injuries have a greater probability of experiencing Post-Traumatic Stress Disorder (PTSD). Victims who lost part of his body functions have felt that they are no longer useful, cannot be productive as it used again and the fear of losing their job. The length of stay in the hospital can also provide its own stress to the victims. Injuries from traffic accidents are the leading cause of death and disability (inability) in general, especially in developing countries (World Health Organization, 2013).

Data injuries caused by traffic accidents still localized and hospital-based (Emergency Section), case reports at the scene of the Traffic Police and the Department of Transportation. There has been no population-based data is an injury to the community and national level. On that basis, the necessary evidence-based baseline data so that it can be used for prevention programs (Riyadina, Suhardi, & Permana, 2009).

6 Affiliations

M. Sc. Nadia Hanum
Institution: University of Leipzig, Clinical Psychology and Psychotherapy
Address: Neumarkt 9–19, 04109 Leipzig, Germany
E-mail: dhea_nadia@yahoo.com

Prof. Dr. Konrad Reschke
Institution: University of Leipzig, Clinical Psychology and Psychotherapy
Address: Neumarkt 9–19, D-04109 Leipzig, Germany
E-mail: reschke@uni-leipzig.de

7 References

American Psychiatric Association. (2000). *The Diagnostic and Statistical Manual of Mental Disorders, 4th edition, Text Revision.* Washington, DC: Author.

Bryant, B., Mayou, R., Wiggs, L., Ehlers, A., & Stores, G. (2004). Psychological consequences of road traffic accidents for children and their mothers. *Psychological Medicine, 34,* 335–346.

European Transport Safety Council. (2007). *Social and Economic Consequence of Road Traffic Injury in Europe.* Brussels.

Harvey, A. G., & Bryant, R. A. (2000). Two-year prospective evaluation of the relationship between acute stress disorder and posttraumatic stress disorder following mild traumatic brain injury. *American Journal of psychiatry, 157,* 626–628.

Levine, Peter. (1998). *Waking the Tiger.* California: North Atlantic Books.

Ministry of Transportation and Ministry of Environment, Republic of Indonesia. (2009). *Indonesian Country Report on Environmentally Sustainable Transport Implementation.* Bangkok: 4th EST Forum.

Nielsen, M. S. (2003). Prevalence of posttraumatic stress disorder in persons with spinal cord injuries: the mediating effect of social support. *Rehabilitation Psychology, 48*(4), 289–295.

Reason, J. T., Manstead, A. S. R., Stradling, S., Baxter, J., & Campbell, K., (1990). Errors and violations on the roads. *Ergonomics, 33,* 1315–1332.

Richmond, T. S., & Kauder, D (2000). Predictors of psychological distress following serious injury. *Journal of Traumatic Stress, 13,* 681–692.

Riyadina, W., Suhardi, M, & Permana, M. (2009). The Pattern and Sociodemographic Determinant of Traffic Injury in Indonesia. *Majalah Kedokteran Indonesia, 59,* 464–472.

Soehodho, S. (2009). *Road Accidents in Indonesia.* Jakarta: Director of Center for Transport Studies, University of Indonesia

Starr, A. J., Smith, W. R., Frawley, W. H., Borer, D. S., Morgan, S. J., Reinert, C. M., & Mendoza-Welch, M. (2004). Symptoms of posttraumatic stress disorder after orthopaedic trauma. *The Journal of Bone and Joint Surgery, 86*(6), 1115–1120.

Ursano, R. J., Fullerton, C. S., Epstein, R. S., Crowley, B., Vance, K., Kao, T.-C., & Baum, A. (1999). Peritraumatic dissociation and posttraumatic stress disorder following motor vehicle accidents. *American Journal of Psychiatry, 156,* 1808–1810.

Zatzick, F. Douglas. (2002). Posttraumatic Stress, Problem Drinking, and Functional Outcomes After Injury. *Archives of Surgery, 137*(2), 200–205.

World Health Organisation (WHO). (2013). *Global Status Report on Road Safety 2013: supporting a decade of action* (Official report). Geneva, Switzerland: Author.

Dian Sari Utami[1,2] & Guangshu Gu[3]

[1] Islamic University of Indonesia, Indonesia

[2] University of Leipzig, Germany

[3] Wuhan Textile University, China

Families in Trauma: Potential Problems and Determinant Factors of Parent-Child Relationship in China and Indonesia

Abstract. Family environment has a strong influence on its members, both parents and children, but especially for children, as the impact of parent-child relationships will influence their future life. Currently in China and Indonesia, many problems arise in parent-child relationships which have negative effects on children's mental health. Crises in families due to lack of positive parent-child relationships impact children's emotions, personality, learning, and social behavior. The aim of this study was to describe the potential problems and protective factors in parent-child relationships in families who have experienced trauma in Indonesia and China based on a culturally informed perspective. Literature review was used as a method to understand the protective factors in parent-child relationships. Building family values orientation (e.g., trust) and enhancing non-violent communication are suggested as primary prevention to help to create positive parent-child relationships, family resilience, and enhanced family well-being.

Keywords: parent-child relationships, parenting problems, risk factors, protective factors, family values.

1 Introduction

Families at risk are currently an issue in countries with high population, density and demographical changes such as China and Indonesia. In 2013, according to World Bank's Data, China had the largest population in the world and Indonesia the fourth largest. The population numbers in both countries have increased significantly every year. Thus, some policies regarding family are applied differently in each country. In order to reduce the rapidly growing population, China's government applies a one-child policy, whereas in Indonesia the family-planning policy has been in place for a long time. It was assumed that controlling population growth would reduce some problems related to families and social issues.

The impact is stronger under stressful or even traumatic conditions due to psychosocial challenges, rapid social changes, and uncertain conditions facing families today, which have negative influences on parent-child relationships (Walsh, 1996). Those uncertain conditions facing families may also cause trauma, such as disaster, war, abuse, violence, chronic illness, disability, poverty, and imprisonment, any of which can negatively impact a family. Further, the lack of a positive parent-child relationship affects children's emotional, personality, educational, and social behavior.

The aim of this article is to describe protective factors in parent-child relationships among families at risk in China and Indonesia. The present chapter first describes the cause of parent-child relationship problems in China and Indonesia. Second, it is necessary to explain the impact of parent-child relationships problems on children. Third, further explanation of protective factors is suggested in order to pursue family well-being.

2 Literature

2.1 Parenting problems and families at risk in China and Indonesia

Problems occur in the parent-child relationship, as when the role of parenting the children is inflexible. Over time, formerly asymmetric parent-child relationships are transformed due to the differences in resources to cope with the environmental and developmental demands of children and parents. On one hand, parents need to fulfill their own needs, while children also need to fulfill their own. However, particularly in parent-child relationship problems, the greatest impact on children is from the parents themselves. Thus, for children, the connection to the parents is considered relevant for the individuation process (Cooper, Grotevant, & Condon, 1983).

Parent-child relationships develop within the family system and will be affected by the wider socioeconomic and cultural context (Schaie & Willis, 1995). Communication problems have been found in parent-child relations in China, which tend to be less demanding, less attentive, and stricter (Shek, 2000). Meanwhile, in Indonesia, due to the excessive focus on material needs, the attention of parents to the fulfillment of children's needs for emotional and spiritual bonding decreases (Makarim, 2008). Problems in parent-child relationships will have negative impacts, especially for children, such as painful stress, emotional problems, academic problems, and behavioral problems (e.g., delinquency). Child abuse, violence, delinquency, and child neglect are found in Indonesian families as reported by Indonesian Commission of Children Protection (KPAI, 2015). In

China, school stress among primary school children (Hasketh et al., 2011), child maltreatment (Fang et al., 2015), and violence have recently become problems in families (UNICEF, 2015).

2.2 Culture and family values in China and Indonesia

Cross-cultural studies have shown that different cultures have different understandings and meanings of parenting (Trommsdorff & Kornadt, 2003). In China and Indonesia, the interaction and relationships between parents and children within families have a strong influence on the life span development of the children. In traditional Chinese culture, the parent-child relationship has to be very close due to cognation (Fong, Yee, & Pattie, 2005). In China, the emphasis shifts from the promotion of independence, sociability, and assertiveness to obedience, compliance, and collectivistic spirit (Harkness & Super, 2002). Further, self-restraint, obedience, and cooperation are highly valued. In Indonesia, parents also tend to have more parental control and at the same time less conflict and more harmony in the parent-child relationship (Trommsdorff, 1995). In Eastern countries, maturity is achieved when children are willing and able to fulfill the role and responsibilities in the hierarchical structure of families (Chao & Tseng, 2002). This responsibility, for example, includes taking care for siblings or younger children in the family. In parent-child relationships, trust is viewed as the relational foundation between children and parents and becomes the indicator of the quality of relationships (Shek, 2008). Trust itself grows through shared experienced knowledge and communication in the family, based on the parents' knowledge of a child's daily activities and demonstrations of responsibility (Kerr, Sattin, & Trost, 1999).

3 Method

This article focuses on the discussion of parent-child relationships among families at risk in China and Indonesia with school-age children. Issues are identified and research findings formulated based on a review of the relevant literature.

4 Results

Based on previous research and literature review, a number of potential problems, risk factors, and protective factors among families in crisis in China and Indonesia have been identified (Table 1).

Table 1. Parent-child relationship issues

	China	Indonesia
Potential problems	Drug abuse, delinquency, domestic violence, child neglect, stress in schools	Delinquency, violence, child abuse, child neglect
Risk factors	Fear, communication, lack of trust, marital problems, low self-esteem, lack of values, autonomous behavior (Ho, 1986)	Lack of trust, lack of communication, parental stress, marital problems, lack of values (Setiadi, 2006)
Protective factors	Positive communication, interactivity, trust within families, have more time with the children (Lerner & Spanier, 1980), respect the children, family beliefs	Value of "family harmony", positive communication, social support, parental self- efficacy, family beliefs

Vulnerable families tend to be in crisis due to several risk factors. These were found both in China and Indonesia, including lack of trust, lack of values, parent-child communication problems, and marital problems, despite socioeconomic factors. It is important for families at risk to deal with stress, whether the stressors precipitate from the social environment or within families. Building an educational home environment in the relationship between parents and children through values sharing is one alternative protective factor for families. Value, such as trust, will be helpful to enhance the sense of commitment in parent-child relationships over the life span. Furthermore, parents also need to view improving their parenting behaviors as a positive activity, as they have invested much time and effort in nurturing their children. Non-violent communication within families is also the most important factor to establish qualitatively positive parent-child relationships in China and Indonesia.

5 Discussion

Enhancing a positive parent-child relationship has become an important issue both in Indonesia and China. Parent-child relations influence the well-being of children and adolescents (Wenk, Hardesty, Morgan, & Blair, 1994), moral development (Bronstein, Fox, Kamon, & Knolls, 2007), and behavioral deviances of adolescents (Baer, 1999). Trust and non-violent communication are assumed to be protective factors that will be helpful to enhance positive parent-child relationships for Indonesian and Chinese families. For future research, a focus on parent-child relationships in China and Indonesia is important in order to identify their unique qualities. This will increase understanding of the impact of protective factors and offer guidance for relevant programs for families at risk due to experiences such as violence, disaster, disabilities, chronic illness, or trauma in both countries. It will also be important to investigate the cause of potential problems among families at risk who have experienced trauma and the outcomes in regard to family resilience and well-being in China and Indonesia.

6 Affiliations

M.A. Dian Sari Utami
Institution: Islamic University of Indonesia
Address: Jalan Kaliurang kms 14.5, Sleman 55584 Yogyakarta, Indonesia
E-mail: dian.utami@uii.ac.id

M.Sc. Guangshu Gu
Institution: Wuhan Textile University
Address: 1 FangZhi Road, Wuhan, P. R. China 430073
E-mail: guguangshu@163.com

7 References

Baer, J. (1999). Family relationships, parenting behavior, and adolescent deviance in three ethnic groups. *Families in Society, 80*(3), 279–285.

Bronstein, P., Fox, B. J., Kamon, J. L., & Knolls, M. L. (2007). The meaning of good parent-child relationships for Mexican American adolescents. *Journal of Research on Adolescence, 17,* 639–667.

Chao, R. & Tseng, V. (2002). Parenting of Asians. In M. H. Bornstein (Ed.), *Handbook of Parenting: Vol. 4. Social conditions and applied parenting* (2nd ed., pp. 59–93). Mahwah, NJ: Erlbaum.

Cooper, C. R., Grotevant, H. D., & Condon, S. M. (1983). Individuality and connectedness in the family as a context for adolescent identity formation and role taking skill. *New Directions for Child Development, 22,* 43–59.

Fang, X., Fry, D. A., Ji, K., Finkelhor, D., Chen, J., Lannen, P., & Dunne, M. P. (2015). The burden of child maltreatment in China: A systematic review. *Bulletin of the World Health Organization, 93,* 175–185C.

Fong, L., Yee, Y., & Pattie. (2005). A search for new ways of describing parent-child relationships: Voices from principals, teachers, guidance professionals, parents and pupils. *Childhood, 12,* 111–137.

Harkness, S. & Super, C. M. (2002). Culture and parenting. In M. H. Bornstein (Ed.), *Handbook of parenting* (2nd ed.). New Jersey: Erlbaum.

Ho, D. Y. F. (1986). Chinese patterns of socialization: A critical review. In M. H. Bond (Ed.), *The psychology of Chinese people*. New York: Oxford University Press.

Kerr, M., Stattin, H., & Trost, K. (1999). To know you is to trust you: Parent's trust is rooted in child disclosure of information. *Journal of Adolescence, 22,* 737–752.

KPAI (Indonesian Commission of Children Protection). Retrieved May 30, 2015, from http://kpai.go.id.

Lerner, R. M., & Spanier, G. B. (1980). *Adolescent development: A life-span perspective*. New York: McGraw-Hill.

Makarim, I. A. (2008, November). Penanaman nilai dalam keluarga, sekolah, dan masyarakat (Value education in family, school, and society). In Ikatan Psikologi Perkembangan Indonesia (IPPI). *IPPI National Scientific Proceedings*. Paper presented at *Temu Ilmiah Nasional IPPI* (pp. 167–169). Bandung: IPPI.

Schaie, K. W. & Willis, S. L. (1995). Perceived family environments across generations. In V. L. Bengston, K. W. Schaie, & L. M. Burton (Eds.), *Adult intergenerational relations* (pp. 174–209). New York: Springer.

Setiadi, B. N. (2006). Indonesia: Traditional family in a changing society. In J. Georgas, J. W. Berry, F. J. R. van de Vijver, Çiğdem Kağıtçıbaşı, & Y. H. Poortinga (Eds.), *Families across cultures: A 30-Nation Psychological Study* (pp. 370–377). Cambridge: Cambridge University Press.

Shek, D. T. L. (2000). Differences between fathers and mothers in the treatment of, and relationship with, their teenage children: Perceptions of Chinese adolescents. *Adolescence, 35,* 135–146.

Shek, D. T. L. (2008). Predictors of perceived satisfaction with parental control in Chinese adolescents: A 3-year longitudinal study. *Adolescence, 43,* 153–164.

Trommsdorff, G. (1995). Parent-adolescent relations in changing societies: A cross-cultural study. In P. Noack & M. Hofer (Eds.), *Psychological responses to social change: Human development in changing environments* (pp. 189–218). Berlin: De Gruyter.

Trommsdorff, G. & Kornadt, H. J. (2003). Parent-child relations in cross-cultural perspective. In L. Kuczynski (Ed.), *Handbook of dynamics in parent-child relations*. London: Sage.

UNICEF. Retrieved May 30, 2015, from http://www.unicef.cn.

Walsh, F. (1996). The concept of family resilience: crisis and challenge. *Family Process Journal, 35,* 261–281.

Wenk D., Hardesty, C. L., Morgan, C. S., & Blair, S. L. (1994). The influence of parental involvement on the well-being of sons and daughters. *Journal of Marriage and Family, 56,* 229–234.

Chapter 3
Intervention Methods

Konrad Reschke

Universität Leipzig, Deutschland

20 Jahre Traumaforschung an der Leipziger Universität: Zur Entwicklung diagnostischer und therapeutischer Techniken (20 Years of Research on Trauma at the University of Leipzig: Development of Diagnostic and Therapeutic Techniques)

Abstract. There was a long period of trauma research in the Division Clinical Psychology and Psychotherapy of the Institute of Psychology at University of Leipzig from 1994 to 2014. More than 40 scientific publications were published in books, articles and other forms. Own experiences of traumatic events and therapeutic case work with trauma-patients were related to the scientific research of trauma related issues. In this paper one example of own work will be reported: (a) The development of the MPSS and (b) The Imagery Rescripting & Reprocessing Therapy (IRRT) as an approach for type I trauma disorders will be mentioned only. The MPSS is the Modified PTDS Symptom Scale (Falsetti, Resick, Resnick & Kilpatrick, 1993). This scale and the other introduced questionnaires for PTSD evaluate the existence and the severity of the PTSD symptoms.

Zusammenfassung. Es war eine lange Periode der Traumaforschung in der Abteilung Klinische Psychologie und Psychotherapie des Instituts für Psychologie an der Universität Leipzig von 1994 bis 2014. Mehr als 40 wissenschaftliche Publikationen wurden in Büchern, Artikeln und in anderer Form veröffentlicht. Eigene Erfahrungen mit traumatischen Ereignissen und therapeutische Fall-Arbeit mit Trauma-Patienten wurden in die eigene wissenschaftliche Trauma-Forschung integriert. In diesem Beitrag soll es um ein Beispiel eigener Arbeit gehen: (a) Die Entwicklung der MPSS und (b) Imagery Rescripting & Reprocessing Therapie (IRRT) als ein Ansatz für die Typ-I-Trauma-Störungen wird nur erwähnt. Die MPSS ist die modifizierte PTBS Symptom Scale (Falsetti, Resick, Resnick & Kilpatrick, 1993). Diese Skala und die anderen vorgestellten Fragebögen für PTBS bewerten die Existenz und den Schweregrad von PTBS-Symptomen.

Keywords: post-traumatic stress disorders, self-judgement, interviews, evaluation.

1 Psychodiagnostik der PTBS

Eine erfolgreiche Traumatherapie setzt immer eine gründliche Diagnostik voraus. Dazu bietet sich im deutschen und vor allem im anglo-amerikanischen Raum ein umfassendes Repertoire an Instrumenten zur diagnostischen Erfassung der posttraumatischen Belastungsstörungen an. Folgende deutschsprachigen, diagnostischen Interviews und Selbstbeurteilungsverfahren sind u. a. in klinischer Anwendung:

Tabelle 1: Fremd- und Selbstbeurteilungsverfahren in klinischer Anwendung

Abkürzung	Instrument	Autor
	Diagnostische Interviews	
SCID-PTSD	Structured Clinical Interview for DSM-IV	Spitzer & Williams, 1986
CAPS	Clinical Administered PTSD Scale	Blake et al., 1990; 1995
DIPS	Diagnostisches Interview bei psychischen Störungen	Margraf et al., 1994
PTSD-I	PTSD-Interview	Watson, 1990; 1991
DIA-X	Diagnostisches Expertensystem für psychische Störungen	Wittchen et al., 1997
	Selbstbeurteilungsverfahren	
IES	Impact of Event-Skala	Horowitz et al., 1979
IES-R	revidierte Version der Impact of Event-Skala	Weiss & Marmar, 1996; Maercker & Schützwohl, 1997
PTSS-10	PostTraumatic Stress Scale-10	Raphael et al., 1989
PDS	Posttraumatic Diagnostic Scale	Foa et al., 1993
PSS	PTSD Symptom Scale	Foa et al., 1993
MPSS	Modified PTSD Symptom Scale	Falsetti et al., 1993
AFT	Aachener Fragebogen zur Traumaverarbeitung	Flatten & Wälte, 1998
ETI	Essener Trauma-Inventar	Tagay et al., 2004

*Notiz.** Literatur in der Tabelle kann beim Autor erfragt werden.

Bei der Analyse bisheriger Selbstbeurteilungsinstrumente zur Darstellung posttraumatischer Belastungsstörungen zeigte sich, dass diese oftmals die Intensität der Symptome nicht erheben bzw. die Kenntnis eines Traumas voraussetzen. Die Modified PTSD Symptom Scale (MPSS) (Falsetti et al., 1993) erfasst als

Screening-Instrument die Symptomschwere unabhängig von der Traumaart bzw. – anzahl (Spitzer, Abraham, Reschke, & Freyberger, 2001). Die Evaluation und die erste klinische Anwendung der Skala im deutschsprachigen Raum werden in der Arbeit von Abraham (1999) berichtet. Die Ergebnisse dieser Arbeit wurden in der Zeitschrift für Klinische Psychologie und Psychotherapie veröffentlicht (Spitzer et al., 2001).

Neben dieser Screeningskala wurden am Lehrstuhl der Klinischen Psychologie in Leipzig weitere diagnostische Verfahren zur Erkennung von PTBS re-evaluiert. „Eines der am häufigsten eingesetzten Selbsteinschätzungsinstrumente für die Erforschung und Diagnostik der posttraumatischen Belastungsstörung (PTBS) ist die Impact of Event Scale-Revised (IES-R). Dieser Fragebogen erfasst mit drei Subskalen die posttraumatischen Symptombereiche ,Intrusion', ,Vermeidung' und ,Übererregung'. Für die deutsche Version der IES-R und einer speziell aus ihr entwickelten Formel für die Prädiktion der PTBS nach DSM-IV liegen bisher nur wenige Befunde vor" (Jaensch, 2011, S. 4). Die Arbeit von Jaensch (2011) bewertet die Qualität der IES-R unter Beurteilung der Reliabilität, der Validität und der Sensitivität anhand einer großen klinischen Stichprobe.

In Kooperation der Universitäten von Gondar und Leipzig entstand eine Arbeit von Wondie (2012). Die Arbeit befasst sich mit einer wesentlichen Quelle kindlicher Traumatisierungen in Äthiopien, dem sexueller Missbrauch in verschiedenen Erscheinungsformen. Er geschieht hauptsächlich in frühen Ehen, bei Vergewaltigungen und Kinderprostitution. Wondie übersetzte die Children's Impact of Traumatic Events Scale-Revised (CITES-R) ins Amharische und überprüfte die interne Konsistenz und die Konstruktvalidität dieser Version an einer Stichprobe sexuell missbrauchter äthiopischer Mädchen. Neben der psychometrischen Bewertung berichtet Wondie Zusammenhänge des Kontexts mit dem Grad der wahrgenommenen sozialen Unterstützung und der Missbrauchsaufsicht sowie der wahrgenommenen Vulnerabilität. Seine Untersuchung wurde 2013 im Journal of Child & Adolescent Trauma veröffentlicht.

Stöber, Reschke und Krause (2013) revidierten ein diagnostisches Verfahren zur Erfassung des Hoffnungskonzepts. Dieses Konstrukt aus dem Forschungsgebiet der positiven Psychologie erfasst eine wesentliche Ressource zur Bewältigung von Anforderungen und belastenden Ereignissen. Das Hoffnungskonzept stammt von C. R. Snyder (1994), es stellt ein gesundheitspsychologisches theoretisches Modell dar. Stöber diskutiert die Anwendung des Hoffnungskonzeptes für die Beratung und Analyse von Menschen in Krisensituationen und berichtet über die Ergebnisse einer Untersuchung zur messtheoretischen Überprüfung der deutschen Version der Dispositional Hope Scale (Snyder, 1994). Damit wird ein

Bogen von der allgemeinen Konzeptbetrachtung zu dessen Aufbereitung für die erfolgreiche Anwendung in Forschung und Praxis gespannt (Stöber et al., 2013).

2 Die Modified PTSD Symptom Scale (MPSS)

Die am weitesten verbreiteten Instrumente zur Abbildung posttraumatischer Belastungsstörungen sind die Impact of Event Scale (IES) und ihre revidierte Version (IES-R). Diese Verfahren beziehen sich auf ein einziges bekanntes Trauma und erfassen intensitätsunabhängig ausschließlich die Symptomhäufigkeit.

Als Screeningverfahren, d. h., unabhängig von der Traumaart bzw. – anzahl entwickelten Foa und Mitarbeiter (1993) die PTSD Symptom Scale, self-report (PSS-SR; deutsche Übersetzung von Winter, Wenninger & Ehlers 1992), die jedoch auch die Symptomhäufigkeit erfasst. Steil (1997) fügte dieser Version eine 4stufige Skala zur Abbildung der Symptombelastung hinzu, die bisher überwiegend an Unfallopfern validiert wurde. Im amerikanischen Raum wurde ebenfalls eine modifizierte Version mit der Möglichkeit, die Symptomintensität auf einer 5stufigen Skala abzubilden, vorgelegt (Falsetti et al., 1993). Diese sogenannte MPSS wurde in klinischen und nicht-klinischen Populationen eingesetzt und zeigte gute psychometrische Eigenschaften. Ihr weiterer Vorteil liegt darin, dass die Ergebnisse international vergleichbar sind. Nachteilig ist allerdings, dass sich die MPSS auf die letzten 14 Tage vor der Untersuchung bezieht und somit ausschließlich als Screeningverfahren fungieren kann. Der Einsatz der MPSS wird bei Patienten mit unbekannter Trauma-Anamnese empfohlen, wie dies häufig bei klinischen Stichproben der Fall ist (Carlson, 1997; Spitzer u. a. 2001).

„Die MPSS umfasst 17 Items, die mit den Symptomen der PTSD entsprechend der DSM-Konzeption korrespondieren. Jedes Item wird auf einer 4stufigen Häufigkeitsskala (von 0 = überhaupt nicht bis 3 = fast immer) und einer 5stufigen Schweregradskala (0 = überhaupt nicht belastend bis 4 = extrem belastend) bezogen auf die beiden letzten Wochen eingeschätzt. Eine kategoriale Auswertungsmethode entsprechend dem DSM-Algorithmus erfasst die Symptome, bei denen eine Häufigkeit von mindestens 1 angegeben wurde. Alternativ ist eine dimensionale Auswertung möglich, bei welcher die Punktwerte für die Häufigkeits- und Schweregradskala für die Gesamt- und getrennt für die Subskalen Intrusionen, Vermeidung und Übererregung addiert werde" (Spitzer et al., S. 160). Die MPSS wird nachfolgend dargestellt.

3 Evaluation und klinische Anwendung der modifizierten PTBS Symptom Skala (MPSS)

Zielsetzung der Arbeit von Abraham war die Bereitstellung eines Screeningverfahrens, welches die Häufigkeit und den Schweregrad der typischen PTBS-Symptomatik in deutscher Sprache erfasst sowie dessen psychometrische Überprüfung.

Ausgehend von der MPSS (Falsetti, Resick, Resnick, & Kilpatrick, 1993) wurde eine deutsche Adaptation erstellt. Dazu wurde die englischsprachige Originalskala übersetzt und mit der deutschen Version des DSM-IV und anderen Skalen zur Erfassung der PTBS-Symptomatik abgeglichen (Abraham, 1999). Als psychometrische Instrumente zur Validierung der MPSS wurden das DIA-X-Interview (Wittchen & Pfister, 1997), die FDS (Freyberger, Spitzer, & Stieglitz, 1999) sowie die SCL-90 von Derogatis (Franke, 1995) genutzt. Die klinische Stichprobe umfasste 103 überwiegend depressive Patienten im Durchschnittsalter von 40 Jahren (Abraham, 1999).

Die Untersuchung der psychometrischen Eigenschaften der MPSS ergab folgende Resultate:

„Als Maß der Reliabilität wurde die innere Konsistenz (Cronbachs Alpha) berechnet. Diese betrug für die Gesamtskala .89 und für die Subskalen fanden sich folgende Werte: Übererregung = .88; Vermeidung = .94 und Intrusionen = .94.

Zur Überprüfung der Validität wurde die Übereinstimmung der verschiedenen Auswertemodi mit einem Fremdbeurteilungsinstrument ermittelt. Mittels des DIA-X konnte bei 44 (42.7 %) der 103 Patienten eine PTSD diagnostiziert werden. Von diesen berichtete einer über Kriegseinsatz, drei hatten Gewaltanwendung erlebt, acht waren Opfer einer Vergewaltigung geworden, vier berichteten über sexuellen Missbrauch als Kind, einer über eine Naturkatastrophe, drei über schwere Unfälle, einer war in Gefangenschaft gewesen und 23 hatten andere Traumata erlebt (wie z. B. den plötzlichen Tod eines nahen Angehörigen oder die Diagnose einer lebensbedrohlichen Krankheit).

Gemäß der kategorialen Auswertungsmethode wurde bei insgesamt 63 Patienten (62 %) der Verdacht auf eine PTSD geäußert. Die Sensitivität der Skala betrug somit 81.4 %, die Spezifität 52.5 % und die Anzahl der richtig identifizierten Fälle 65 %. Die von den Autoren vorgeschlagene modifizierte kategoriale Auswertung zeigte, dass bei 32 Patienten (31 %) möglicherweise eine PTSD vorlag. Die Sensitivität lag bei 74.4 %, die Spezifität bei 79.6 % und die Gesamtübereinstimmung richtig erkannter Fälle betrug 77.5 %.

Neben der kategorialen Auswertung ermöglicht die MPSS eine dimensionale Auswertung sowohl für das gesamte Verfahren als auch für die Subskalen, wobei Symptomhäufigkeit und -schweregrad als Summenscores einfließen. In einem

varianzanalytischen Vergleich zeigten Patienten, bei denen mittels des DIA-X
eine PTSD diagnostiziert wurde, dabei durchweg signifikant höhere Werte als
diejenigen ohne PTSD, was als weiterer Hinweis auf die Validität gelten kann
(Spitzer et al., 2001).

„Mittels der Regression (Testwert X = 0.107 * Intrusionsscore + 0.0622 * Ver-
meidungsscore – 0.0589 * Übererregung – 2.2883) ließen sich insgesamt 82.4 %
der Patienten diagnostisch richtig zuordnen. Dabei wurde für Patienten mit einem
Testwert X ≥ .5 eine PTSD-Diagnose vergeben. Die Sensitivität gemäß dieser
Regressionsgleichung betrug 76.7 % und die Spezifität 86.4 %" (Abraham, 1999,
S. 33).

4 Fazit

Mit der MPSS steht ein reliables und valides Screeninginstrument für Trauma-
folgestörungen zur Verfügung, welches je nach Aufgabenstellung angepasst ausgewertet werden kann. Steht die Entdeckung möglichst aller Personen mit PTBS
im Vordergrund, empfiehlt sich die Anwendung der dichotomen Auswertungsmethode (Sensitivität = 81,4 %). Allerdings bedarf es vor der Diagnosestellung
einer weiteren spezifischen Diagnostik, da diese Methode zu einer erheblichen
Anzahl falsch positiver Kategorisierungen führt (Spezifität = 52,5 %). Eine relativ
effiziente kategoriale Beurteilung gelingt unter Verwendung der modifizierten
Auswertungsmethode, bei welcher neben der Symptomhäufigkeit die Symptomschwere Eingang findet. Bei diesem Vorgehen sind die diagnostischen Kennwerte
vergleichsweise ausgewogen. Die dimensionale Auswertung steht ergänzend für
andere Fragestellungen zur Verfügung und ermöglicht eine differenzierte Beurteilung bezüglich des Symptomausmaßes zwischen verschiedenen Gruppen.

Die MPSS kann zuverlässige Hinweise auf eine PTBS geben. Es wurde auch gezeigt, dass niedrig-gradige Stressoren zu annähernd so ausgeprägten subjektiven
PTBS-Symptomen wie hochgradige Stressoren führen können. Diese Erkenntnis
widerspricht der bisher verbreiteten Ansicht, dass niedrig-gradige Stressoren eher
selten zu einer PTBS führen. Darüber hinaus ist auf die enge Assoziation zwischen
Traumatisierung und dissoziativer Psychopathologie hinzuweisen.

Zusammenfassend kann festgestellt werden, dass die Befunde der Untersuchung bezüglich der Reliabilität der deutschen Version der MPSS als mindestens
ebenso zufriedenstellend einzuschätzen sind, wie die Reliabilitätskennwerte der
MPSS Originalversion und anderer diagnostischer Instrumente im deutschen
Sprachraum. Die Ergebnisse dieser Diplomarbeit wurden von Spitzer u. a. (2001)
in der Zeitschrift für Klinische Psychologie und Psychotherapie veröffentlicht.

5 Kontakt

Prof. Dr. Konrad Reschke
Institution: University of Leipzig, Clinical Psychology and Psychotherapy
Address: Neumarkt 9–19, 04109 Leipzig, Germany
E-mail: reschke@rz.uni-leipzig.de

6 Literatur

Abraham, G. (1999). *Evaluation und klinische Anwendung der Modifizierten PTSD Symptom Skala (MPSS).* Unveröffentlichte Diplomarbeit, Universität. Leipzig.

Carlson, E. B. (1997). *Trauma assessment. A clinicans guide.* New York: Guilford Press.

Falsetti, S. A., Resnick, H. S., Resick, P. A., & Kilpatrick, D. G. (1993). The Modi-fied PTSD Symptom Scale: a brief self-report measure of posttraumatic stress disorder. *The Behavior Therapist 16*, 161–162.

Franke G. H. (1995). *SCL-90-R. Die Symptom-Check-Liste von Derogatis. Deutsche Version.* Weinheim: Beltz.

Freyberger, H. J., Spitzer, C., & Stieglitz, R.-D. (1999). *Fragebogen zu Dissoziativen-Symptomen.* Bern: Huber.

Jaensch, R. (2011). Evaluation der Impact of Event Skala - revidierte Version. *Unveröffentlichte Diplomarbeit*, Universität Leipzig.

Snyder, C. R. (1994). *The psychology of hope: You can get there from here.* New York: Free Press.

Spitzer, C., Abraham, G., Reschke, K., & Freyberger, H. J. (2001). Die deutsche Version der Modified PTSD Symptom Scale (MPSS): Erste psychometrische Befunde zu einem Screeningverfahren für posttraumatische Symptomatik. *Zeitschrift für Klinische Psychologie und Psychotherapie*, 30(3), 159–163.

Steil, R. (1997). *Posttraumatische Intrusionen nach Verkehrsunfällen.* Frankfurt: Peter Lang.

Stöber, F. S., Reschke, K., & Krause, S. (2013). Das Hoffnungskonzept nach C. R. Snyder: Eine Ressource zur Krisenbewältigung. In E. Witruk & A. Wilcke (Hrsg.). *Historical and cross-cultural aspects of psychology* (pp. 511–527). Frankfurt: Peter Lang.

Winter, H., Wenninger, K., & Ehlers, A. (1992). *Deutsche Übersetzung der PTSD Symptom Scale-self-report (PSS).* Psychologisches Institut der Universität Göttingen.

Wittchen, H.-U., & Pfister, H. (1997). *DIA-X-Interviews: Manual für Screening-Verfahren und Interview.* Frankfurt: Swets & Zeitlinger.

Konrad Reschke

Wondie, Y., Zemene, W., Reschke, K., & Schröder, H. (2012). The psychometric
    properties of the Amharic version of the children's impact of Traumatic Events
    Scale-Revised: A study on child sexual abuse survivors in Ethiopia. *Journal of
    Child & Adolescent Trauma, 5(4)*, 367–378.

Dian Veronika Sakti Kaloeti & Evelin Witruk

University of Leipzig, Germany

Prison Parenting Rehabilitation Programs as a way to Reduce Traumatic Experience Caused by Parental Incarceration

Abstract. Parental incarceration has a detrimental effect on children's lives. Research indicates these children are traumatized by separation from their parents, confused by the parent's actions, experience instability and are stigmatized by the shame of their parent situation. Incarcerated children are also at higher risk of experiencing other traumas such as neglect. It is shown that imprisonment disrupts the positive relationship between parents and their children. Maintaining family contact during incarceration can be beneficial to both children and their parents. Research demonstrates that strengthening positive family connections through rehabilitative parenting programs during incarceration and after incarceration can yield positive societal benefits in the form of reduced recidivism. These programs which aim to enhance parent child relationship, on the same line can also promote of healthy child development. This article aims to present the hardship resulting from parental incarceration and to summarize the effects of maintaining parent-child relationships during incarceration. The brief also highlights key factors that should be considered when developing programs, and provide several recommendations and program strategies for incarcerated parents. The opportunity, and challenging situation regarding Indonesia's prison rehabilitation context is also discussed.

Keywords: incarcerated parent, traumatic experience, prison parenting programs, rehabilitation programs.

1 Introduction

Parent-child relationships begin early in children's lives and are critical for children's long-term adjustment and success. The evidence consistently indicates that parent-child relationships are linked to children's development, adjustment, well-being, and educational attainment throughout the life course (Thornton, Orbuch, & Axinn, 1995). Therefore, parenting becomes an important theme with no exception for incarcerated families.

Over the last decade, much research has been conducted showing the negative effects of parental incarceration upon children (Arditti, 2005; Johnston, 1995; Johnson, 2009; Luke, 2002; Mumola, 2000; Murray & Farrington, 2008; Parke

& Clarke-Stewart, 2001; Trice & Brewster, 2004), and for this reason, they are regarded as vulnerable population. The degree of negative impact is affected by various factors, including the role the parent played in the child's life prior to incarceration as well as the age of the child at the time incarceration occurs (Parke & Clarke-Stewart, 2001). The impact of parental incarceration on families also has been conceptualized as a form of family crises.

Maintaining the relationship while inside prison is a challenging situation for the families. Parenting programs for incarcerated parents will help to overcome the issues and improve incarcerated parent rehabilitation. Programs can help parents reconnect and maintain a relationship with their children as well as improving their parenting skills. These programs can also provide services to other families' members in supporting incarcerated parents transition back into the community and reuniting with their children successfully (Harris, Graham, & Carpenter, 2010).

2 Theory

Parenting issues among the incarcerated is a unique parenting experience. There are some issues regarding their condition such as difficulties in arranging and maintaining contact with the children. This is a source of stress for incarcerated parents (Kaloeti, 2011). Maintaining family contact during incarceration can be beneficial to both children and their parents, possibly through personal visit activities during incarceration. With regards to incarcerated parents, several studies found that the maintenance of family ties during incarceration is linked to post-release success, defined as lower rates of recidivism and fewer parole violations (Hairston, 2003).

Whilst parents are imprisoned, most children go, or continue, to live with the other parent or relatives (Mumola, 2000), and they are called as the co-parent. The co-parent is defined as the children's current caregiver. The relationship between incarcerated parent and the co-parent is becoming one of the important factors that affect whether a child has contact with an incarcerated parent. The communication between incarcerated parents and their co-parent or families provides the most concrete and visible strategy that families and prisoners use to manage separation and maintain connections.

When children do not live with a parent, visits are essential to maintain their relationship. Visiting as a consequence of parent-child separation is seen as routine in the context of a parent's incarceration. Families that visit their imprisoned relatives allow co-parent, incarcerated parent, and children to share family experiences and help them to remain emotionally attached. These visits allow children to express their emotions regarding the separation. While the incarcerated parent can feel reassured that their children have not forgotten them and that they continue to function in socially acceptable roles.

Strengthening family bonds through parent education programs is an effective correctional rehabilitation strategy. Prison programming for parents has the goal of improving outcomes for prisoners and their children, both during and after incarceration. Specifically, they aim to define their personal parenting values, roles and parenting goals to facilitate effective relationships with their children. They can be empower parents to use newly learned skills with their children. The programs could support them in their efforts for effective relationships while incarcerated and after they are released. Despite having similar goals, rehabilitation programs in terms of parenting programs in prison differ significantly in design, execution, and method of assessment. Typical parenting curricula include education regarding effective parenting techniques and child development; the other components may include enhanced visiting, parental rights training, nursery programs, or support groups. There are several types of parental rehabilitation program for incarcerated parents:

a. *Parent education courses:* programs that are tailored to parent learner needs while incorporating essential parenting elements. Providing parenting classes to educate the parents in their parenting role, enhance their child rearing skills, and adapt to the new family roles. Strengthening the parenting skills of prisoners and to provide an opportunity for bonding between incarcerated parents and their children. This programs giving chance for parent-child visitation where to be expected as positive interaction.
b. *A nursery program:* a program that allows incarcerated mothers to keep their babies near them while they are incarcerated. Prison in to Indonesia permit incarcerated mothers to take care of the babies inside a prison until 2 years old.
c. *Parent support groups:* A program that is similar to self-help organization. These groups meet on a regular basis and allow incarcerated parents to explore a wide variety of topics.

3 Discussion

Addressing the specific issues and barriers that incarcerated parent face in parenting are essential in developing rehabilitation program. Specifically, there are challenges in implementing programs based on Indonesia's prison rehabilitation context. Establishing family services in prison is challenging due to hurdles regarding the level of financial and institutional support required. As well as a lack of resources such as educational staff, mental health staff, volunteers from religious or community groups, or even incarcerated parents themselves. Incorporating visitation as part of any rehabilitation program is difficult where many prison environments discourage visits from children and their parents. The facilities also do not acomodate children's needs during visitation and facilities may be regarded as child-unfriendly. There is also a lack of dependable and affordable public transportation and often the prisons are located a long distance away. Thus it is a hardship on children and family members to visit the incarcerated parents. The prisoners's sentence lengths also have an effect on the implementation of the programs. Offenders with long term sentences need support in sustaining connections to their children, while inmates with short-term sentences need skills in maintaining an active parenting role and communicating with their children's caretaker to ease their transition into the family unit once released.

Family support programs must be regarded as essential components of release preparation rather than as programs that reward criminal behavior. It is therefore vital to prepare resources effectively and thoroughly, for example: enhance correctional staff skills with specific training. Programs should adopt a systemic approach, developing relationships with schools, correctional facilities, and other community agencies, all working together to implement the program. After that, designing curricula and programs that are relevant to the prisoner population and specific cultural and ethic norms. In sum, using local wisdom about parenting values, is suggested as more appropriate approach for developing prison parenting rehabilitation programs within Indonesian population.

4 Affiliations

M.Psi. Dian Veronika Sakti Kaloeti
Institution: University of Leipzig, Educational and Rehabilitation Psychology
Address: Neumarkt 9–19, 04109 Leipzig, Germany
E-mail: veronikasakti@gmail.com

Prof. Dr. Evelin Witruk
Institution: University of Leipzig, Educational and Rehabilitation Psychology
Address: Neumarkt 9–19, 04109 Leipzig, Germany
E-mail: witruk@uni-leipzig.de

5 References

Arditti, J. A. (2005). Families and incarceration: An ecological approach. *Families in Society, 86*, 251–258.

Hairston, C. F. (2003). Prisoners and their families: parenting issues during incarceration. In J. Travis, & M. Waul (Eds.), *Prisoners Once Removed: The Impact of Incarcerations and Reentry on Children, Families, and Communities* (pp. 260–282). Wahington DC: The Urban Institute.

Harris, Y. R., Graham, J. A., & Carpenter, G. J. O. (2010). *Children of incarcerated parents: theoretical, developmental, and clinical issues.* New York: Springer Publishing Company.

Johnston, D. (1995). Effects of parental incarceration. In K. Gabel, & D. Johnston (Eds.), *Children of Incarcerated Parents* (pp. 59–88). New York: Lexington Books.

Johnson, R. (2009). Ever-increasing levels of parental incarceration and the consequences for children. In S. Raphael, & M. Stoll (Eds.), *Do prison make us safer? The benefits and costs of the prison boom* (pp. 177–206). New York: Russel Sage.

Kaloeti, D. V. S. (2011). *Prisoners and their children: Parent child relationship issues behind bars (perspective from Indonesia).* In W. Srisayekti (Ed), Proceedings of Padjadjaran International Conference on Psychology: "Psychology for a better future", October 23–26, 2011. Bandung, Indonesia: Faculty of Psychology Padjadjaran University.

Luke, K. P. (2002). Mitigating the ill effects of maternal incarceration on women in prison and their children. *Child Welfare, 81*, 929–948.

Mumola, C. (2000). *Incarcerated parents and their children.* Washington DC: U.S. Department of Justice, Office of Justice Programs.

Murray, J., & Farrington, D. (2008). The effects of parental imprisonment on children. In M. Tonry (Eds), *Crime and Justice: A Review of Research* (pp. 136–206). Chicago, IL: University of Chicago Press.

Parke, R. D., & Clarke-Stewart, K. A. (2001). *Effects of parental incarceration on young children.* Prepared for the "From Prison to Home: The Effect of Incarceration and Reentry on Children, Families, and Communities Conference", January 30–31, 2002. U.S. Department of Health and Human Services and The Urban Institute. Retrieved from http://aspe.hhs.gov/sites/default/files/pdf/74976/report.pdf.

Thornton, A., Orbuch, T. L., & Axinn, W. (1995). Parent-child relationships during the transition to adulthood. *Journal of Family Issues, 16*(5), 538–564.

Trice, A. D., & Brewster, J. (2004). The effects of maternal incarceration on adolescent children. *Journal of Police and Criminal Psychology, 19*(1), 27–35.

Yumi Lee[1], Yun-Hee Kim[2], Ji-Hye Kang[3], & Hyeong-Keun Yu[4]

[1] University of Leipzig, Germany

[2] Hannam University, South-Korea

[3,4] Korea National University of Education, South-Korea

Postvention is Prevention: Helping Students Bereaved by Suicide in Korean Schools

Abstract. People bereaved by suicide (i.e., suicide survivors) often are close to the person who committed suicide. Suicide survivors comprehend these accidents as traumatic experiences. They experience difficulties in accepting the committed suicide either through denial or by keeping themselves in silence. In particular, students in the adolescent period are the most highly affected by their friends' behavior, thus they opt to imitate their friend's suicide. Therefore, keeping an eye on these suicide survivors and also providing them with postvention programs are critical in order for them to cope with the suicide incident effectively and eventually go back to daily life. The purpose of this study was to develop a school postvention program in order to prevent another suicide in schools in South-Korea.

Key Words: suicide survivors, Bronfenbrenner, traumatic experiences, counselor, teacher.

1 Suicide in South-Korea

Korea's[1] suicide rate was the highest in the OECD countries for 12 years, from 2002 to 2013 (OECD, 2013). According to Statistics Korea data, 28.5 suicides per 100,000 people (39.5 people per day). Youth suicide is also a major concern in Korea. The suicide rate among adolescents (age: 10–19) showed that suicide accounted for 29.5 % (6.5 adolescents per day) which is the first most frequent cause of death among adolescents in Korea. In 2010, the Korean Ministry of Education reported the suicide rate of school students and in total 735 (elementary school: 17, middle school, 224, high school, 494).

In 2010, Lee, Hong, and Espelage investigated the youth suicide phenomenon based on Bronfenbrenner (1994)'s ecological system theory in order to understand within a larger context (e.g., individual, families, schools, peers, and communities)

---

1 From here on, Korea represents South-Korea

(see *Figure* 1). They found that Korean youth suicide can be best understood in a larger environmental context: (a) *characteristic levels*: depression, hopelessness, impulsivity, substance use, internet addiction, and academic stress: (b) *micro-system*: parent-child relationships, parent-child communication, peer support, peer victimization, and school satisfaction, (c) *meso-system*: interrelationship between more than two micro-system (e.g., relationship between family and school), (e) *exo-system*: mass media and suicide internet sites, (f) *macro-system*: collectivistic culture and emphasis on academic achievement, and (g) *chrono-system*: previous major financial crisis in 1997 which broke down the country's social layer.

Figure 1. The summary of the Korean Youth Suicide Phenomenon based on Bronfenbrenner (1994)'s the Ecological System Theory (Lee et al., 2010)

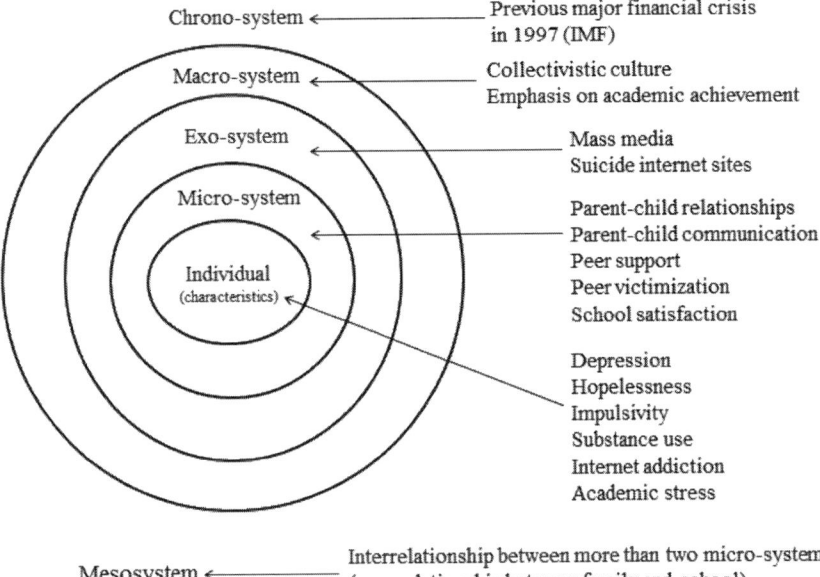

2 Suicide survivors

Suicide is an individual action. However, its consequential negative effect is considerably large. In general, people bereaved by suicide (so called suicide survivors) often are close to the person who committed suicide. Suicide survivors comprehend these accidents as traumatic experiences. They experience difficulties in accepting the committed suicide either through denial or by keeping themselves in silence. Furthermore, suicide survivors feel guilty about not being able to prevent the suicide, so they either blame or isolate themselves socially. In particular, students in the adolescent period are the most highly affected by their friends' behavior, thus they opt to imitate their friend's suicide. Therefore, keeping an eye on these suicide survivors and also providing them with post-vention programs are critical in order for them to cope with the suicide incident effectively and eventually go back to daily life. The purpose of this study was to develop a school postvention program to provide school counselors, homeroom teachers, parents, and school administrators guide lines in order to reduce the rate of another suicide in Korean schools.

3 Postvention model for suicide survivors in Korean schools

Figure 2 show the postvention model for suicide survivors in Korean school, involving school counselors, home room teachers, parents, and school adminis-trators. Specific role of each of them is as follows:

Figure 2. Postvention Model for Suicide Survivors in Korean School

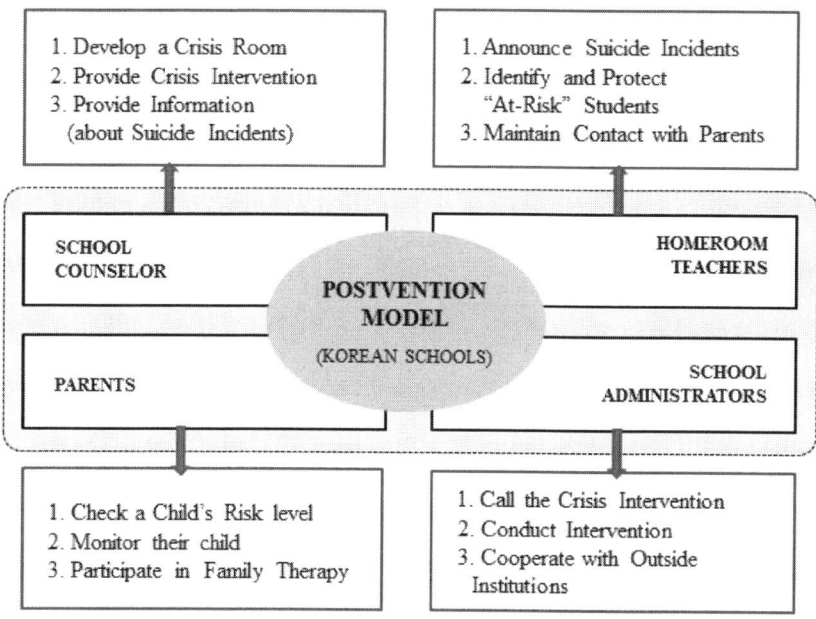

3.1 School counselors

The first step of postvention programs is helping the suicide survivors to accept the reality of the suicide by expressing and sharing negative feelings through crisis counseling sessions. For this reason, the school counselor needs to be aware of the suicide survivors' behaviors in order to help them to overcome their friends' death. Postvention programs need to be planned and organized in advance, conducted by well-trained school counselors, and provided immediately after the committed suicide. In addition, school counselors should be able to recognize and assess the potential risk of suicide survivors as they are the students, who need long-term counseling,

3.2 Homeroom teachers

Suicide is the most chaotic situation for most of the students. If committed suicide happened in a class, homeroom teachers should take care of the students who became suicide survivors. For this reason, the homeroom teacher needs to acknowledge that postvention programs are the most effective and productive prevention to help suicide survivors in order to cope with at-risk situations.

3.3 Parents

If your child becomes a suicide survivor, parents need to monitor their child's psychological conditions and behavior attentively. The child feels guilty especially when he/ she is close with the students who committed suicide. Therefore, parents need to check their child's current conditions by consistently monitoring and participating family therapy.

3.4 School administrators

Suicide affects the entire school. Almost students (and staffs) experience intensive psychological trauma when a suicide is committed. In this situation, the school needs to conduct a well-organized systemic postvention program so that suicide survivors feel stable and are able to adjust to their daily life accordingly.

4 Discussion

Figure 3. Postvention is Prevention

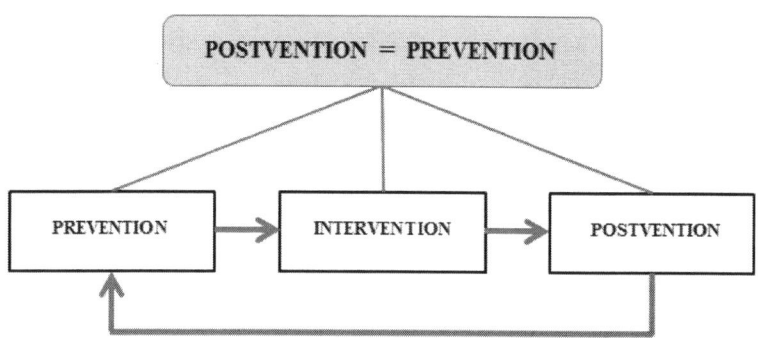

Postvention is Prevention! Postvention can be key to preventing another suicide in Korean schools (see *Figure* 3). The current postvention model (based on the role) could not only help school staff (i.e., school counselors, teachers, and school administrators) but also parents in order to help suicide survivors and to prevent another suicide in schools in Korea.

5 Affiliations

Dr. Yumi Lee
Institution: University of Leipzig, Educational and Rehabilitation Psychology
Address: Neumarkt 9–19, 04109 Leipzig, Germany
E-mail: yumi.lee@uni-leipzig.de

M.A. Yun-Hee Kim
Institution: Hannam University, Counseling
Address: 70 Hannamro, Daedeok-gu, Daejeon 34430, South-Korea
E-mail: helper201205@gmail.com

M.A. Ji-Hye Kang
Institution: Korean National University of Education, Counseling and Special Education
Address: 250 Taeseongtabyeon-ro, Grangnae-myeon, Heungdeok-gu, Cheongju-si, Chungbuk 28173, South-Korea
E-mail: dawn0829@hanmail.net

Prof. Dr. Hyeong-Keun Yu
Institution: Korean National University of Education, Counseling and Special Education
Address: 250 Taeseongtabyeon-ro, Grangnae-myeon, Heungdeok-gu, Cheongju-si, Chungbuk 28173, South-Korea
E-mail: yhkcem87@knue.kr

6 References

Bronfenbrenner, U. (1994). Ecological models of human development. In T. Husen, & T. N. Postlethwaite (Eds.), *The international encyclopedia of education* (2nd ed., pp. 1643–1647). New York: Elsevier Science.

Korean Ministry of Education. (n.d.) Retrieved from http://english.moe.go.kr/enMain.do.

OECD (2013). *"Suicides", in OECD Factbook 2013: Economic, Environmental and Social Statistics*, OECD Publishing. Retrieved from http://www.oecd.org/els/health-systems/MMHC-Country-Press-Note-Korea.pdf.

Lee, S.-Y., Hong, J.-S., & Espelage, D. L. (2010). An ecological understanding of youth suicide in South Korea. *School Psychology International, 31* (5), 531–546.

Yu, H.-K., Kim, Y.-H., Kang, J.-H., & Lee, Y. (2012). *Suicide in Children and Adolescents in Korea* [in Korean]. Seoul, Korea: Hakjisa.

Marcus Stueck

DPFA University of Applied Sciences Saxony, Germany

Ten Steps of Stress Reduction: The Intercultural Adapted Version of Training of Stress Reduction with Elements of Relaxation (STRAIMY®-International)

Abstract. In this article, the author explains the intercultural adapted version of the training with elements of relaxation. In 10 Steps there is explained how is the way of coping stress: psycho-educative pathway: (1) Stress-Definition, (2) Situation, (3) Reaction, (4) Attitude, (5) Consequences, (6) Coping, (7) Resources, (8–9) Wishes and Aims (10) Closing. The emotional-regulative pathway are lead through exercises (e.g., Yoga, Islamic or Buddhist praying, Autogenic Training), especially focused for the use in international context (e.g., Latvia, Indonesia, Nepal, Iran) and indicated for treatment of stress disorders and stressful traumatic circumstances (e.g., natural disasters such as Nepal-earthquake, Merapi-eruption Indonesia or car accidental stress in Iran and Germany). The second part of the article introduces some results of evaluation studies of STRAIMY in Germany and Latvia. Because of its scientific basic, the training is listed in the German Psychological Association as an accepted evidence based stress reduction training.

Keywords: stress reduction with relaxation, intercultural evidence based program stress reduction, evidence based Yoga program, STRAIMY.

1 Introduction

In this article, the author explains the actual version of the training with elements of relaxation (e.g., Yoga, Islamic or Buddhist praying, Autogenic Training). This is especially focused for the use in international context (e.g., Latvia, Indonesia, Nepal, and Iran) and indicated for treatment of stress disorders and stressful traumatic circumstances (e.g., Nepal-earthquake, Merapi-eruption in Indonesia or car accidental stress in Iran or Germany).

Concerning stress reduction there are two ways of auto regulation internal coping (reflexive activities & emotional regulation) and two ways of external coping (changing conditions & learning abilities to solve problems). Based on this Model of Lazarus and Schroeder (as cited in Stueck, 2008) the following steps (sessions) are designed in 2 parts: 1: **Psycho-education** (external and internal coping) and

2: **Emotional regulation** (Relaxation techniques, Islamic or Buddhist praying, Meditation, Yoga)

Psycho-education	Internal coping (reflexive activities, exercises focused on behavioral concept of S-O-R-K (Situation, Organism, Reaction, Consequences) external coping (abilities concerning stress evoking condition)
Relaxation	Yoga, Breathing-Meditation, Buddhist Meditation, Islam praying

The STRAIMY®-Training was developed by the author of this article at the Institute of Psychology, University of Leipzig from 1999 to 2008. Since then, many STRAIMY®-instructors all over the world were educated (Latvia, Germany, Nepal).

In ten training sessions (or in three days/compact or seminar) stress reduction related competencies based on latest psychotherapeutic insights. They are conducted in the following areas:

- individual diagnosis of stress exposure and stress signals and their health consequences,
- building skills to manage short-and long-term exposures better,
- dealing with unpleasant emotions and feelings, time management,
- working on the individual goal and desire formation.

The importance of this training as a stress coping possibility was proved, as part of a comprehensive scientific monitoring-investigation (see point 3).

2 Ten steps of stress reduction

These ten steps are related to the actual use of the exercises in the STRAIMY® and to the handbooks of the original version. They have been proofed to work most in the practice and you will find the pages in the handbooks.

Figure 1. Logo and Handbook of STRAIMY®

The handbooks in English is available at http://www.bildungsgesundheit.de. The German version of the handbooks can be found at http://www.schibri.de.

Table 1.   Ten Steps (Sessions) of Stress Reduction (the Pages are Related to the Handbook)

Sequences		Homework
Psycho-education **Step 1**	Introduction "Relation", Group Rules Information: What is stress? Definition of stress	Observe situation that makes stress. Relaxation e.g., yoga, breathing (concentrate on your breathing if you feel nervous)
Emotional-Regulation	Relaxation, Yoga, Breathing-Meditation, Praying	
Psycho-education **Step 2**	Conclusion stressor analysis Information: Stress reduction start in your head Information: How leads stress to illness? First we do OBSERVATION and UNDERSTANDING and ACCEPTANCE S – O – R – C – Structure	Observation of stressors, Relaxation e.g., exercise yoga, breathing, praying
Emotional-Regulation	Stressor-analysis Relaxation, Yoga, Breathing-Meditation, Praying	
Psycho-education **Step 3**	Autogenic Training Conclusion stressor analysis Stressor – Reaction – Analysis If you describe situations, you have to describe it very concrete. Please describe exactly your Reactions (Body, Feeling, Thinking, Behavior)	Observations of Stress Situations and their reactions, relaxation e.g., exercise yoga, breathing, praying
Emotional-Regulation	Relaxation, Yoga, Breathing-Meditation, Praying	
Psycho-education **Step 4**	Teaching of Autogenic Training Conclusion: Stressor – Reaction – Analysis Organism variable **Attitudes** which push stress-reaction **Needs:** 3 existential questions: 1) Where do I want to live? 2) What do I want to do? 3) With whom do I want to live? **Biology:** Vegetative working point (pulse, blood pressure)	Identify stress pushing attitudes, Observation of your blood pressure relaxation e.g., exercise yoga, breathing, praying
Emotional-Regulation	Relaxation, Yoga, Breathing-Meditation, Praying	

Sequences		Homework
Psycho-education **Step 5** Emotional-Regulation	Autogenic Training Conclusion: Attitudes – analysis Consequences – analysis (with AVEM/Work-related Experience and Behavior Pattern, psychosomatic chronic reactions) Relaxation, Yoga, Breathing-Meditation, Praying	Reading about AVEM-patterns and interpretation, relaxation e.g., exercise yoga, breathing, praying
Psycho-education **Step 6** Emotional-Regulation	Autogenic Training Conclusion Consequences Stress-coping "Lazarus Exercise" Example: Anger-regulation (anger in, anger control, anger out; short-long-term techniques) STEPS: Express anger by yourself, accept state afterwards, e.g., sadness, than speak with other ("do not shoot someone if you are angry") Relaxation, Yoga, Breathing-Meditation, Praying	3 questions of Lazarus, techniques of anger regulation, Relaxation, Yoga, Breathing-Meditation, Praying
Psycho-education **Step 7** Emotional-Regulation	Autogenic Training Resources: social support (e.g., social emergency hand, social atoms) Friendship Exercise of Enjoyment Relaxation, Yoga, Breathing-Meditation, Praying	Visit your friends Relaxation, Yoga, Breathing-Meditation, Praying
Psycho-education **Step 8** Emotional-Regulation	Autogenic Training Exercise: How I want to use my lifetime? Wishes, IGEL-list Relaxation, Yoga, Breathing-Meditation, Praying	IGEL-List Relaxation, Yoga, Breathing-Meditation, Praying
Psycho-education **Step 9** Emotional-Regulation	Autogenic Training Treasure map of wishes Relaxation, Yoga, Breathing-Meditation, Praying	Treasure map of wishes Relaxation, Yoga, Breathing-Meditation, Praying
Psycho-education **Step 10** Emotional-Regulation	AIMS Repeating, conclusion Scale of AIMS Relaxation, Yoga, Breathing-Meditation, Praying	Transformation of the training aims (sent by the STRAIMY-instructor) Relaxation, Yoga, Breathing-Meditation, Praying

3 STRAIMY-training compact or special additional modules

There are special options for additional modules or special applications of training:

a) After three months, there will be refreshing workshop for STRAIMY-Course participants (analyzing of the STRAIMY-Aims).
b) For working with trauma-victims or helpers of natural catastrophes (i.e., earthquake, high water, car accidents), there will be 1 or 2 more sessions. The training has up to 15 to 20 sessions and depends on the impact of traumatization. Maybe accompanying individual EMDR sessions are necessary.
c) There are three days training, e.g., for companies:

Day 1: Session 1–3 of Psycho-educational Part, Relaxation Part (e.g., Yoga or other techniques, Buddhist, Islam Praying)
Day 2: Session 4–6 of Psycho-educational Part, Relaxation Part (e.g., Yoga or other techniques, Buddhist, Islam Praying)
Day 3: Session 7–10 of Psycho-educational Part, Relaxation Part (e.g., Yoga or other techniques, Buddhist, Islam Praying).

4 Scientific evaluation

STRAIMY concept and efficiency has been studied in various projects and diploma papers and results of the studies have been summarized in Stueck's habilitation work, which was worked out from 1999 to 2007. Also at Educational university in Riga there was done a study with STRAIMY (Svence, Kazaga 2009). In Leipzig, a study was carried out on STRAIMY efficiency for teachers (Stueck, Rigotti, & Mohr, 2004).

Results showed improvement of psychological (feeling of comfort, ability to relax), physiological (relaxing effect, skin sensitivity, pulse) and immunological (immunoglobulin A) of teachers involved in the study. An instant and long-lasting decrease of consequences caused by stress was detected in the sense of a more successful functioning of a personality (communication models in working environment, self-efficiency, ability to regulate emotions, control of anger improved) (Stueck, Meyer, Rigotti, Bauer, & Sack, 2003; Stueck, 2008). The results showed also a decrease of sympathetic activation in the 24-hour-monitoring between the first and the second measurement (pre, post) of the yoga training. A high effect size ($d' > 0.8$) and very high power ($1\text{-}\beta > 0.8$) were shown with a long-term shift of the vegetative area from sympathetic states of strain to increasing parasympathetic states of relaxation. Moreover, a better activity-relaxation-synchronization of basal-rest-activity-cycles (BRAC) was also be observed in this study as a result of STRAIMY. This means that the test persons easier reach their ideal ratio of strain and relaxation with 67 % activation patterns and 33 % relaxation patterns after both inventions that is an indicator for health (Stueck, Villegas, Perche, & Balzer, 2007).

5 Discussion

Overall, there is a huge impact on the effects of STRAIMY according to the psycho-physiological health state of the tested persons. The different researches can prove the positive effects and show the range of this special intervention method. Doing this specific process in getting knowledge according to stress and health in combination with the use of Yoga-elements is a great mixture for improving the health situation of anyone. So far, there is no real contraindication according to the STRAIMY-method, but there is a big need to stay in contact with the persons during the intervention process to focus on their individual needs.

6 Affiliation

Prof. Dr. Marcus Stueck Institution: DPFA Hochschule Sachsen, University of Applied Sciences Saxony, Germany
Address: Mahlmannstrasse 1, 04109 Leipzig, Germany
E-mail: marcus.stueck@dpfa-hs.de

7 References

Stueck, M. (2007). *Entwicklung und empirische Ueberpuefung eines Belastungsbewaeltigungskonzepts für den Lehrerberuf.* (Habilitation). Fakultaet fuer Biowissenschaften, Pharmazie, und Psychologie der Universitaet Leipzig, Leipzig.

Stueck, M. (2008). Neue Wege: Yoga und Biodanza in der Stressreduktion mit Lehrern. In M. Stueck (Ed.), *Neue Wege in Paedagogik und Psychologie, Bd. 1.* Strasburg: Schibri-Verlag.

Stueck, M., Meyer, K., Rigotti, Th., Bauer, K., & Sack, U. (2003). Evaluation of a Yoga-based stress management training for teachers: Effects on Immunoglobulin A secretion and subjective relaxation. *Journal of Meditation and Meditation-Research,* 2(1), 59–68.

Stueck, M., Rigotti, Th. & Mohr, G. (2004). Untersuchung der Wirksamkeit eines Belastungsbewaeltigungstrainings fuer den Lehrerberuf. *Psychologie in Erziehung und Unterricht,* 51(3), 236–245.

Stueck, M., Villegas, A., Perche, F., & Balzer, H.-U. (2007). Neue Wege zum Stressabbau im Lehrerberuf: Biodanza und Yoga als koerperorientierte Verfahren zur Reduktion psycho-vegetativer Spannungszustaende. *ErgoMed, 03*, 68–75.

Svence, G. & Kazaka, T. (2009). Sress Reduction Programme Straimy and Biofeedback Measurement System Adaption in Riga Teacher Training and Educational Management Academy. *Journal of Pedagogy and Psychology "Signum Temporis"*, *2*(1), 114–123.

Edgar Galindo

Universidade de Evora, Portugal

An Intervention Program for Children with School Failure Problems

Abstract. School failure is a chronic problem in many developing countries and also in some developed ones as Portugal due to factors like cognitive deficiencies of children, an inadequate family environment, low SES and bad teaching methods or school organization. The problem persists in spite of many applied measures. It is important to develop effective and simple intervention strategies which are able to cope with the problem at an individual level independently of the cause. The general objective of the following study is to develop cognitive-behavioural training techniques to help children with problems of school failure. 6–12 years old children attending ISCED 1, with academic difficulties because of family problems, social exclusion or poor schooling were trained. Cognitive-behavioral techniques that have been widely used to train persons with intellectual, sensorial, physical or social deficiencies were applied. Results are evaluated in terms of % of attained objectives, time and (subjective) teacher satisfaction. Some results of individual children are shown. Training programs seem to be successful independently of the (mostly unknown) cause of failure or the trained school subject.

Keywords: cognitive-behavioural techniques, school failure, disadvantaged children.

1 Introduction

School failure is a main problem in Portugal and in many other countries. Data published by the European Union (Eurydice, 2011) show that Portugal had the highest retention rate at primary school (ISCED 1) in 2001–2008, but Spain and France have also got the problem. These and similar data are the first signs of an urgent need for intervention at ISCED 1 in order to look for a solution. School failure has often been defined emphasizing the misbehaviour of the child. Faubert (2012) offers a more operational definition: "…school failure, (is) understood as the failure of schools and the school system to provide the appropriate level of, and adequately defined services for, all students to be successful" (p. 4). Common manifestations of school failure are leaving school before ending education, successive failure leaving to a décalage between school year and the chronological age of children (rate of retention) and referring children to special education. Educational agencies have proposed several causes, like cognitive deficiencies of children, an inadequate family environment, low SES, bad teaching methods or schools. Consequently, several strategies have been applied, like more pre-school education, adapting school to children needs, better teachers, technicians & teaching methods, better relationship between school and social environment and family (Faubert, 2012). They are all probably right, because a real solution implies an intervention at all levels. Nevertheless, the problem persists. One main contribution of Psychology to solve the problem is developing differential teaching methods for individual children. Now, in treating an individual case, the cause of school failure is not the most important point; indeed, most of the time you can only suppose a set of possible causes. At an individual level, it is important to develop effective and simple intervention strategies to cope with the problem independently of the cause. The general objective of this study has been to develop cognitive-behavioural diagnostic and training techniques to help children with school failure problems. The publication of handbooks for teachers and parents with simple and efficient programs is a main goal. School success was defined in terms of a set of skills proposed by teachers; school failure was then understood as a lack of one or more of these skills. Consequently, the specific objectives of this study are: 1) to develop intervention techniques for ISCED 1 based on school defined skills, 2) to apply systematically training programs to attain those skills, and 3) to publish these techniques.

2 Literature

Cognitive-behavioural techniques have been applied widely to treat persons with different disorders. Training children with intellectual, sensorial physical or sensorial deficiencies has already a long history. A set of intervention tools to train a large variety of skills, especially in children, has been developed (see Miltenberger, 2012), including specific learning disorders (Levingston, Neef, & Cihon, 2009). School failure problems have been analyzed by Adelman and Taylor (1993), recommending the use of behavioural techniques; Hallahan, Kauffman and Lloyd (1999) and Wallace, Larsen, and Elksnin (1992) identified several causes of school failure, pointing at the absence of language, social and cognitive previous skills (precurrents), as an obstacle to learn reading, writing and mathematics. The important role of precurrents has been confirmed by Carroll, Snowling, Hulme and Stevenson (2003), Connor, Son, Hindman, and Morrison (2005) and Leppänen, Niemi, Aunola, and Nurmi (2004). Last but not least, Galindo, Galguera, Taracena, and Hinojosa (2009) have applied cognitive-behavioural programs in Mexico to train slum children with intellectual, sensorial, or social deficiencies; this is the basis of the work done in Portugal, explained in this study.

3 Method

3.1 Participants and setting

Participants in this study were low SES 6–12 years old children attending ISCED 1 (1st, 2nd and 3rd school year), referred by teachers because of school failure of different origins (family problems, social exclusion or poor schooling). Especially trained Psychology students worked as tutors on a 1x1 basis. Each tutor became responsible for evaluating and training a child during a semester (15 weeks). Until now, this intervention program has had two phases: A first phase was carried on in Lisbon with 13 children of low income families (2006–2007) and 80 children of Cova da Moura, a slum-like area of Lisbon, inhabited mainly by immigrants (2007–2008); a second phase is presently going on in Alentejo, an underdeveloped rural region in Portugal, where 57 children have been treated since 2013.

3.2 Instruments and materials (training programs)

Each training program has roughly the same structure: 1) A general objective and a set of behaviourally defined specific objectives, 2) a definition of precurrent skills, 3) training phases and/or steps, 4) setting, 5) procedures (shaping, model training, instructions, etc.), 6) quantitative evaluation (% of correct responses and/or attained objectives), 7) motivational aspects. The following programs have been developed as a basis for further adaptation to each individual case: 1) Basic Behaviour (Precurrents): Self Care, Motor Coordination, Discrimination of Forms & Colors, Pre-reading, Pre-Writing, Verbal Behaviour, Temporal & Spatial Relationships. 2) Academic Behaviour (1st, 2nd & 3rd school years): Reading, Writing, Environment, Mathematics, and Portuguese. 3) Social Behaviour.

3.3 Procedure

Each child was evaluated individually, looking of the existing and failing skills, according to the learning aims defined by his/her teacher; the core of the evaluation is a direct behavioural observation, but an analysis of existing reports on the child is also included. Evaluation proceeded as far as possible, trying to identify lacking precurrents to academic skills. A list of problems was made for each case; on this basis, a hierarchy of intervention aims was defined and an individually tailored program for each skill was designed. In the first phase, training programs were applied 4 hours a week during at least one semester on an individual basis (1 tutor x 1 child); in the second phase, time was reduced to two hours per week. A token economy was introduced (Ayllon & Azrin, 1968). The level of success of each training program was evaluated in terms of a) % of attained objectives (or correct responses to a program), b) training time, and c) (subjective) level of satisfaction of the teacher. A child was due to attain at least 80 % of objectives (or correct responses) in order to be considered successful. As a result of this individualized procedure, the nature, duration and amount of trained programs greatly varied from child to child.

4 Some results

A handbook containing intervention programs and results of the first phase is ready for publication (Galindo, in press). Therefore, I will show as an example some results related to the second phase, emphasizing that the main interest is to show the effects of training in each individual child. *Figure* 1 shows the results obtained by four children in school year 2013–1014, independently of the trained subject. The percentage of correct responses obtained in a pre-test (in black) and a post-test (in grey) is shown. In all cases, there is an improvement in the post-test compared with the pre-test. Child A had improvements in two subjects (30 % to 90 % and 60 % to 90 %); child B advanced from 80 % to 100 % in the only trained subject; child C shows mixed results in two subjects, from 16 % to 50 % in one (good, but still not successful) and from 30 % to 87 % in the second; child D showed great improvements in general, from 30 % to 89 %, from 72 % to 100 %, from 13 % to 78 % (almost successful) and even from 0 % to 100 % of correct responses.

Figure 1. Individual Results Obtained in Some Training Programs by Four Children

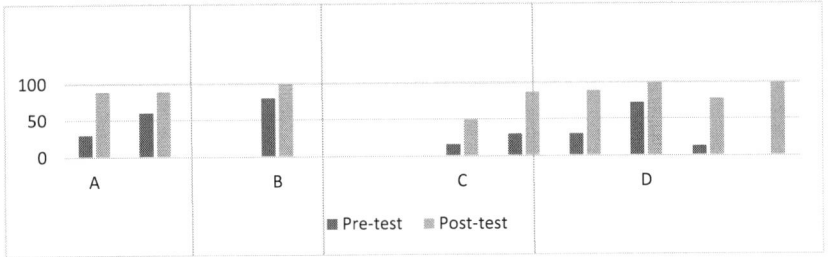

5 Disscussion

Results show an apparent improvement of academic behaviour in all cases, as a probable consequence of intervention. Similar results have been reported by authors treating school failure with cognitive-behavioural techniques (Carroll et al., 2003; Connor et al., 2005; Leppänen et al., 2004). Teachers reported significant positive changes in all cases, but sometimes they complained a child had improved its behaviour but still had academic problems, because some children were trained in precurrent and/or social behaviours rather than in reading & writing. These results seem to show applied programs are successful to train a set of skills, whose absence may cause school failure problems. Notwithstanding, more research is needed with other children, skills, ages, etc., in order to make a sound contribution to the solution of school failure at an individual level.

6 Affiliation

Prof. Dr. Edgar Galindo
Institution: Universidade de Evora
Address: Departamento de Psicologia, 7005–345 Evora, Portugal
E-mail: ecota@uevora.pt

7 References

Adelman, H. S., & Taylor, L. (1993). *Learning problems and learning disabilities: moving forward.* Pacific Grove, C.A: Brooks/Cole Publishing Company.

Ayllon, T. & Azrin, N. H. (1968). *The token economy: A motivational system for therapy and rehabilitation.* Englewood Cliffs: Prentice Hall.

Carroll, J. M., Snowling, M. J., Hulme, C., & Stevenson, J. (2003). The development of phonological awareness in preschool children. *Developmental Psychology, 39*(5), 913–923.

Connor, C. M., Son, S., Hindman, A. H., & Morrison, F. J. (2005). Teacher qualifications, classroom practices, family characteristics and preschool experience: Complex effects on first graders vocabulary and early reading outcomes. *Journal of School Psychology, 43*(4), 343–375.

Eurydice (2011). *Grade retention during compulsory education in Europe. Regulations and statistics.* Brussels: Eurydice.

Faubert, B. (2012). *A literature review of school practices to overcome school failure.* OECD Education Working Papers, 68. Paris: OECD Publishing.

Galindo, E. (In press). *O tratamento do insucesso escolar com técnicas da psicologia* (Treatment of school failure with psychological techniques). Lisboa: Livros Horizonte.

Galindo, E., Galguera, M. I., Taracena, E., & Hinojosa, G. (2009). *La modificación de conducta en la educación especial* (Behavior modification in special education) (2nd rev. ed.). Mexico: Editorial Trillas.

Hallahan, D., Kauffman, J., & Lloyd, J. W. (1999). *Introduction to learning disabilities.* Boston: Allyn & Bacon.

Leppänen, U., Niemi, P., Aunola, K., Nurmi, J. E. (2004). Development of reading skills among preschool and primary school pupils. *Reading Research Quarterly, 39*(1), 72–93

Levingston, H. B., Neef, N. A., & Cihon, T. M. (2009). The effects of teaching precurrent behaviors on children's solution of multiplication and division word problems. *Journal of Applied Behavior Analysis, 42*(2), 361–367.

Miltenberger, R. G. (2012). *Behavior modification* (5th ed.). Wadsworth: Belmont S.A.

Wallace, G., Larsen, S., & Elksnin, L. (1992). *Educational assessment of learning problems*. Boston: Allyn & Bacon.

Chapter 4
Qualitative Research Methods

Bodo Krause

Humboldt-University of Berlin, Germany

Methodische Entwicklungen in der qualitativen Persönlichkeitsforschung (Method Development in Qualitative Personality Research)

Abstract: Scientific research comes increasingly into question about the reliability of qualitative findings in Personality Psychology. Beside some fallacies, it arises also the deficiencies of methodological investigation as antecedents. This is illustrated by three examples within the scope of the complex conclusion in the qualitative research of personality; it will be discussed and evaluated. The starting point is the unit of building reasonable hypotheses, appropriate study design (survey), feasible data evaluation and analysis, including the adequate interpretation of the findings. These methodological requirements on empirical studies, involving in three complex topics: the causes and developments of factor models, the item-response models, and the reliability assessment. In all ranges, it would be new, advance accesses developed, which also differentiate the capture of the phenomenon range, facilitate further development of theoretical background, and bring about practical usage.

Zusammenfassung. Zunehmend stellen wissenschaftliche Recherchen die Verlässlichkeit insbesondere qualitativer Befunde in der Persönlichkeitsforschung infrage. Neben Täuschungen stellen sich vor allem methodische Unzulänglichkeiten der Untersuchung als Ursachen heraus. Dies wird an drei Beispielen dargestellt und im Rahmen der komplexen Zusammenhänge in der qualitativen Persönlichkeitsforschung diskutiert und bewertet. Ausgangspunkt ist die Einheit von begründeter Hypothesenbildung, angemessener Versuchsplanung (Untersuchung) und zulässiger Datenauswertung und –analyse sowie der adäquaten Interpretation der Befunde. Dies begründet methodische Anforderungen an empirische Untersuchungen, deren Entwicklungen wir an drei Themenkomplexen nachvollziehen: den Ursachen und Entwicklungen der Faktorenmodelle, der Itemantwortmodelle und der Beurteilung der Zuverlässigkeit. In allen Bereichen wurden neue, erweiterte Zugänge entwickelt, die sowohl eine differenziertere Erfassung des Phänomenbereichs ermöglichen als auch eine Weiterentwicklung der theoretischen Hintergründe und praktischen Anwendungen bewirken.

Keywords: Modellierung der Beobachtungssituation, Faktorenmodelle, Itemantwortmodelle.

1 Ausgangspunkt

In der Fachliteratur finden wir wiederholt Darstellungen und Diskussionen, die die Verlässlichkeit und Allgemeingültigkeit psychologischer Aussagen und Theorien infrage stellen. Drei Beispiele sollen dies verdeutlichen:

1.1 Ein klassisches Beispiel sind die Befunde aus der Zwillingsforschung zur Vererbbarkeit der Intelligenz, wie sie von Cyril Burt vorgelegt wurden, der für seine Erkenntnisse anschließend geadelt wurde. Der Spiegel berichtet schon 1978, dass eine Analyse von Donald D. Dorfman beweist, dass die grundlegenden Daten dieser Studie manipuliert und gefälscht sind. „Mit dem Denkmal Burt stürzen auch die Dogmen, die sich aus seinen Forschungsergebnissen begründeten: Reiche sind klüger als Arme, Schwarze sind dümmer als Weiße". Und weiter wird ausgeführt: „Nur eine Zunft, vermutet das amerikanische Wissenschaftsblatt „Science" könnte die Zahlenfälschereien des Psychologen Burt vielleicht erklären: die Psychologen."

Aus forschungsmethodischer Sicht entspricht die Forderung nach Reproduktion/ Replikation wissenschaftlicher Befunde zum einen der Prüfung der Zuverlässigkeit dieser Befunde und andererseits wird sie Gegenstand von metanalytischen Vorgehensweisen (z. B. Publikationsbias, file-drawer-analysis).

1.2 Ein aktuelles Beispiel berichtet Konitzer (2013) unter dem Titel „Einmal ist keinmal" mit dem Untertitel „Immer wieder wird die Aussagekraft psychologischer Studien angezweifelt. Tatsächlich sind etliche nicht reproduzierbar, wie aktuelle Beispiele zeigen."

Ausgangspunkt sind Studien des niederländischen Sozialpsychologen Diederik Stapel, die zeigen dass

- Fleischesser ungeselliger und egoistischer als Vegetarier sind,
- Straßen voller Müll die Diskriminierung fördern,
- Frauen, die nach der Heirat den Namen des Mannes annehmen, für weniger intelligent gehalten werden,
- Frauen, die nach der Heirat ihren Namen behalten, als ehrgeiziger und weniger fürsorglich gelten.

„Inzwischen steht fest: All das ist Humbuk. Viele Jahre lang hatte der Soziologe diese und andere Ergebnisse gefälscht."

1.3 Ein weiteres, aktuelles Beispiel finden wir unter dem Titel „Big Five unter Beschuss" (Tenzer, 2013). Mit Bezug auf die Ergebnisse eines Forscherteams um den Anthropologen Michael Curven resümiert sie, dass offensichtlich das Big-Five-Modell nicht generell für Menschen zutreffend ist. Aus den Untersuchungen folgt u. a., dass Menschen in abgelegenen Dörfern in Bolivien nur zwei Faktoren aufweisen (Fleiß und soziale Nützlichkeit) während Menschen in Südafrika neun Faktoren als persönlichkeitskonstituierend aufweisen. Auch Menschen in China sind mit dem Big-Five-Modell nicht kompatibel. Entscheidend dabei ist, dass es nicht nur um die Anzahl der Faktoren geht, sondern dass es unterschiedliche, nicht übereinstimmende Faktoren sind. Im Beispiel von Bolivien sind die beiden Faktoren Fleiß und soziale Nützlichkeit natürlich eng mit dem Lebensumfeld und dem Ziel der Arterhaltung verbunden.

2 Methodischer Ausgangspunkt

Anliegen der Persönlichkeitsforschung ist es allgemein, den Zusammenhang von Persönlichkeitseigenschaften und Umweltbedingungen hinsichtlich der Handlungs-regulation zu untersuchen. Menschliches Erleben und Verhalten wird untersucht, seine Veränderungen in ihren Ursachen aufgeklärt.

Für unsere Betrachtungen sind dabei drei Ausgangspunkte wesentlich:

a) Die Ursachen der Verhaltensregulation sind der externen Beobachtung häufig nicht zugänglich. Sie werden als latente Variable (auch Faktoren, Konstrukte) bezeichnet und sollen aus dem beobachtbaren Verhalten erschlossen werden (manifeste Variable, Beobachtungsmerkmale, Messwerte). Der klassische Ansatz ist die Black-box-Methode, die versucht deren Zusammenhang aufzuklären.

b) Jede einzelne Beobachtung an einer Person ist zufallsabhängig (das Be-obachtungsergebnis ist nicht mit Sicherheit vorhersagbar) und in dieser Form i.R. auch nicht wiederholbar.

c) Die wissenschaftliche Untersuchung dieser Zusammenhänge macht die Erfassung der Ausprägungsgrade psychischer Eigenschaften erforderlich, also die Messung psychischer Eigenschaften. Dies ist im Bereich der Persönlichkeitseigenschaften häufig dadurch erschwert, dass die Merkmale keine Metrik aufweisen, sondern als Qualitäten auftreten. (Letztere können zusätzlich eine Ordnung aufweisen).

Abbildung 1. Gegenstand indirekter Messmethoden

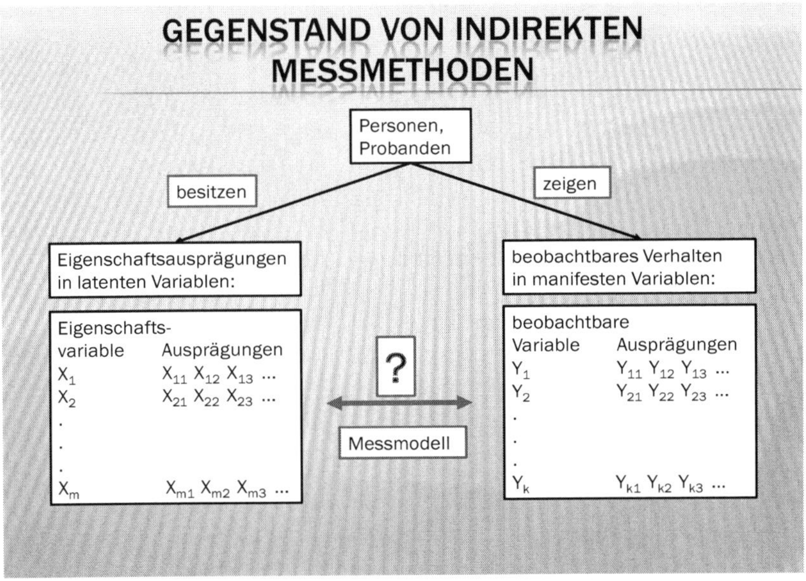

Zur Erfassung von Persönlichkeitseigenschaften sind also sowohl geeignete Mess-
modelle oder Maßstäbe als auch geeignete Prüf- und Entscheidungsmethoden
erforderlich. Gigerenzer (1981) ergänzt, dass auch das Subjekt, die Zielsetzung
und der Gegenstandsbereich einbezogen werden müssen. Dies kennzeichnet die
Komplexität der Untersuchungsmethodik und soll nachfolgend in seinen wesentlichen Entwicklungen dargestellt werden. Eine solche Messsituation hatten wir
schon früher (Krause, 2013) durch das folgende Schema dargestellt:

Einen ersten fundamentalen Zugang für ein solches Messmodell entwickelte
Carl Friedrich Gauß (*30. April 1777 in Braunschweig, † 23. Februar 1855 in
Göttingen) im Rahmen seiner umfassenden Arbeiten zur Vermessung der Erde.

In seinem Buch „Gauß und die Messkunst" kennzeichnet Lelgemann (2011)
den universellen Zugang von Gauß mit der folgenden Darstellung (S. 26): „Seit
Gauß bildet die Zerlegung von Messdaten in vier Bestandteile die Grundlage jeder
sachgerechten Auswertung von Messdaten.

$$l_i = \{f_i(x_j,p_j) + s_i\} + \{\varepsilon_i(y_k,q_k) + g_i\}$$

wobei gilt:
- $f_i(x_j,p_j)$ = geometrisch-physikalisches Modell
- s_i = systematischer Fehler

- $\varepsilon_i(y_k, q_k)$ = zufälliges-stochastisches Modell
- g_i = grober Fehler
- p_j, q_k = bekannte Größen
- x_j, y_k = zu berechnende Größen".

Die Betrachtungen machen deutlich, dass es für die Bewertung und Eignung von Messmodellen neben der Frage nach der Existenz und Eindeutigkeit noch folgende weitere prinzipielle Fragekomplexe gibt:

a) die Fragen nach der *Prüfbarkeit* der Voraussetzungen der Modelle (und damit auch nach ihrer Existenz),
b) die Fragen nach der *Bedeutsamkeit* der Beziehungen zwischen den latenten und manifesten Variablen,
c) die Fragen nach den *auftretenden Resten* sowie der *Zuverlässigkeit und Gültigkeit* der Messung.

Zur letzten Frage ist anzumerken, dass auch hier schon Gauß den entscheidenden Zugang formulierte. In der oben zitierten Darstellung von Lelgemann (2011) heißt es: „Bereits 1789 erfand Gauß dort zur Auswertung von Messdaten die „Methode der kleinsten Residuen" Später entwickelte er diese Methode in offensichtlich lang andauernder Arbeit und Erfahrung mit realen Messdaten weiter zur „Methode der kleinsten Quadrate", wozu er erstmalig die sog. zufälligen Fehler durch ein stochasti-sches mathematisches Modell erfaßte, durch die sog. Gaußsche Normalverteilung-Funktion" (S. 26).

Im Rahmen psychologischer Untersuchungen ist die Klassische Testtheorie (KTT) ein typisches Messmodell, das mit seinem personenorientierten Ansatz

$$X_i = t + E_i$$

(mit der Zufallsgrößen X_i für die Beobachtungen, t dem wahren Wert einer Person (true score) und E_i als Fehlerzufallsgröße) den obigen Anforderungen entspricht. Der true score *t* wird als Erwartungswert aus unabhängig wiederholten Beobachtungen durch den Mittelwert abgeschätzt. Bezogen auf eine Untersuchungspopulation wird auch der wahre Wert eine Zufallsgröße und es ergibt sich der allgemeine Modellansatz

$$X = T + E.$$

Die wahren Werte ergeben sich regressionsanalytisch, wobei ein zusätzlicher Bezug zur Reliabilität entsteht:

$$T = \rho^2(X,T) \cdot X + [1 - \rho^2(X,T)] \cdot E(X).$$

Dieses Modell enthält keinerlei freie Parameter (!!), kann also nicht an Daten angepasst werden. Entweder sind seine Voraussetzungen erfüllt, dann ist es zutreffend, oder nicht. Im letzteren Fall wäre zu beurteilen, ob Abweichungen mit der Robustheit des Modells noch verträglich sind (wir kommen später darauf zurück).

Eine Erweiterung erfährt die KTT 1904 durch die Arbeiten von Carl Spearman, der damit einen universellen Zugang zur Messung der Intelligenz begründet:

$$X_i = a_i \cdot g + b_i \cdot s_i$$

wobei Intelligenzleistungen durch das Wirken eines Generalfaktors g (interpretiert als ein Maß der allgemeinen und angeborenen „geistigen Energie") und einem (variablenspezifischen) Restfaktor s_i erklärt werden (Zwei-Faktoren-Theorie oder Generalfaktor-Theorie der Intelligenz). Dieser Ansatz enthält mit den Faktorladungen a_i und b_i freie Parameter, über die das Modell an die Daten angepasst wird. Für die Beurteilung der Angemessenheit dieses Ansatzes formulierte schon Spearman das Tetradendifferenzkriterium (vgl. Überla, 1968, S. 142/143), das jedoch im Weiteren nicht bestätigt werden konnte. Damit ist eine Allgemeingültigkeit dieser Theorie nicht gegeben. Dennoch war dies der Ausgangspunkt für die Weiterentwicklung zur Gruppenfaktorentheorie der Intelligenz, die durch Louis Leon Thurstone (1931) zum Modell der multiplen Faktorenanalyse erweitert wurde. Mit diesem Ansatz konnte Thurstone (1938) sieben relativ unabhängige Faktoren geistiger Fähigkeiten (primary mental abilities) ausweisen (Primär-Faktoren-Theorie der Intelligenz): Zahlenrechnen (numbers), Sprachverständnis (verbal comprehension), Raumvorstellung (space), Gedächtnis (memory), schlussfolgerndes Denken (reasoning), Wortflüssigkeit (word fluency) und Auffassungsgeschwindigkeit (perceptual speed).

Ihre Gültigkeit wird durch das Auftreten schwacher Interkorrelationen getrübt, die als Hinweis auf einen Generalfaktor angesehen wurden. In der diagnostischen Praxis wurden auf dieser Basis u. a. das Leistungsprüfsystem L-P-S von Horn (1983) und der Intelligenzstrukturtest IST von Amthauer (1953, 1973) entwickelt.

Diese Entwicklung verdeutlichen auch, dass ein Modell nicht a-priori für einen Gegenstandsbereich oder eine Zielstellung angemessen sein muss. Vielmehr sind im Sinne der drei obigen Fragekomplexe die Beurteilung eines Messmodells erforderlich und daraus ggfs. Weiterentwicklung dieses Ansatzes zu begründen. Solche Methodenentwicklungen vollziehen sich in Wechselwirkung mit psychologischen Theorien und liefern so einen Beitrag zum Erkenntnisgewinn in der Psychologie. Am Beispiel der obigen linearen Modelle der Intelligenz ist dies z. B. die implizite Annahme der Unabhängigkeit der Faktoren, die eine orthogonale Einfachstruktur bilden sollen. Experimentell ausgewiesene Abhängigkeiten zwischen den Faktoren sind der Ausgangspunkt für schiefwinklige Lösungsansätze, die ein hierarchisches Faktorenmodell zu begründen gestattet (vgl. Guilford). Solche Weiterentwicklungen von Methoden, Modellen und Theorien wollen wir auch an weiteren Methodenentwicklungen speziell in der qualitativen Persönlichkeitsforschung aufzeigen.

3 Methodische Entwicklungen

Methodischen Entwicklungen kennzeichnet Rost (2003) allgemein dadurch, dass sie nicht dazu führen, „dass sich das bisherige Arsenal in Forschungsmethoden als falsch oder unbrauchbar erweist. Vielmehr führen sie dazu, dass sich das Methodenarsenal um wesentliche Teile erweitert und bereichert." (S. 1).

Solche Entwicklungen wollen wir in drei Bereichen nachvollziehen:

3.1 in den Faktormodellen
3.2 in den Itemantwortmodellen (IRT, item response model)
3.3 in der Beurteilung der Zuverlässigkeit (Reliabilität)

3.1 Itemantwortmodelle Faktorenmodelle entwickelten sich aus dem Modellansatz der KTT u. a. zur multiplen Faktorentheorie (vgl. oben L. L. Thurstone). Diese setzte zunächst unabhängige Faktoren und voneinander und den Faktoren unabhängige Restfaktoren voraus:

$$X_i = \sum a_{ij} F_j + a_i R_i$$

Die Faktorladungen sind die freien Modellparameter, die im Rahmen der Faktorenanalyse (z. B. als Modellanpassung an eine orthogonale Einfachstruktur) geschätzt werden. Die Güte einer Faktorenlösung wurde zunächst durch die Varianzaufklärung der gemeinsamen Faktoren (communality) gekennzeichnet. Später dann auch durch den Vergleich der Originaldaten mit vom angepassten Modell her reproduzierten Daten (Chi-Quadrat Homogenitätstest) beurteilt. Eine entscheidende Frage blieb jedoch, ob die damit implizierte orthogonale Einfachstruktur den Beobachtungsdaten angemessen ist. Für die Beurteilung dieser Hypothese entwickelte Bargmann (1954) einen Signifikanztest und bei deren Ablehnung war der Übergang zur schiefwinkligen Rotation, also einem hierarchischen Modell mit voneinander abhängigen Faktoren angezeigt:

$$X_i = \sum a_{ij} F_j + \sum b_{ij} F_j F_j + a_i R_i$$

Die Abhängigkeiten zwischen den Faktoren begründeten dann den Übergang zu einer nachfolgenden zweiten Faktorenanalyse (hierarchisches Modell) zu deren Aufklärung. Damit waren für Faktormodelle die ursprünglichen Einschränkungen auf Eindimensionalität und Unabhängigkeit der Faktoren überwunden. Eine weitere Voraussetzung blieb aber erhalten: die Unabhängigkeit der Fehler untereinander

und von den Faktoren. Ihre Überwindung führte zu einer sehr allgemeinen Modellklasse, den Strukturgleichungsmodellen (SEM, structural equation modelling). Mo-
delliert werden die Zusammenhänge zwischen latenten und manifesten Variablen,
wobei zusätzlich freie Parameter (Gewichte, Varianzen, Covarianzen) einbezogen
werden können. Die Ansätze der Pfadanalyse und der Fehler-in-den-Variablen
Modelle sind Spezialfälle von SEM. Ein einfaches Strukturgleichungsmodell zeigt
die Abbildung 2. Es besteht aus zwei Messmodellen für die latenten und manifesten
Variablen und dem vermittelnden Strukturgleichungssystem:

Abbildung 2. Strukturgleichungsmodell

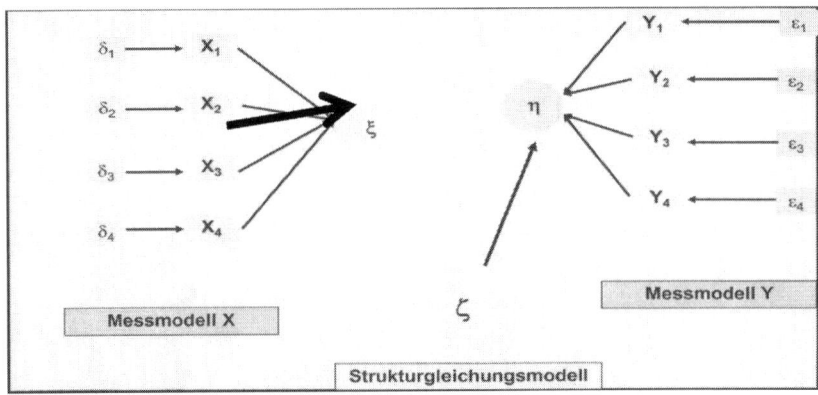

Wesentlich wird nun der Zugang: Diese Modelle enthalten unterschiedlich viele
freie Parameter, die an die Daten angepasst werden. Die Anpassungsgüte wird
durch unterschiedliche Anpassungstests ausweisbar. Wird die Nullhypothese der
Passung nicht verworfen, liegt ein „passendes" Modell vor. Davon kann es aber
mehrere geben. Es gibt im Rahmen des Signifikanztest ja prinzipiell keine Möglichkeit, die Passung zu beweisen. Alternative Modelle können nur mit Bezug
auf den theoretischen Hintergrund diskutiert und bewertet werden (explorative
Untersuchungsstrategie).

3.2 Itemantwortmodelle (IRT) sind eine zweite wichtige Modellklasse, weil sie
direkt für qualitative Daten ausgelegt sind. Ihr Prototyp ist durch das von Rasch
(1960) begründete Prinzip einer stochastischen Itemantwortfunktion gekennzeichnet, die das dichotome Antwortverhalten (0, 1) einer Person j in Abhän-
gigkeit von ihrer Fähigkeit Fj und der Schwierigkeit einer Aufgabe si beschreibt:

$$P\{(a_{ij})=1\}=(\exp(f_j-s_i))/(1+\exp(f_j-s_i)) \quad P\{(a_{ij})=0\}=1-P\{(a_{ij})=1\}=1/(1+\exp(f_j-s_i)).$$

Dieses Modell hängt direkt mit der Eigenschaft zusammen, dass der Summenscore eine erschöpfende Statistik ist, also alle eigenschaftsrelevante Information enthält. Rasch konnte zeigen, dass diese Eigenschaft eine notwendige und hinreichende Bedingung dafür ist, dass es genau ein Messmodell gibt, die obige logistische Funktion. Sie enthält mit der Schwierigkeit genau einen variablenspezifischen freien Parameter, der an die Daten angepasst wird.

Interessanterweise ist diese logistische Funktion (LV) einer Normalverteilung (NV) sehr ähnlich. Lord und Novick (1968, S. 399) zeigen diese Beziehung:

$$|NV(x/1,7)-LV(x)|<0,01 \text{ für alle } x.$$

Zusätzlich besteht die Möglichkeit, die Passfähigkeit des Raschmodells durch einen Anpassungstests zu prüfen und bei seiner Nichtablehnung von der Gültigkeit auszugehen. Damit wird dann der Summenscore als erschöpfende Statistik einer quasi-normalverteilten Grundgesamtheit ausgewiesen. Schon Rost (1999) wies darauf hin, dass dies der entscheidende Fortschritt ist, der Rechtfertigkeitsnachweis für die Verwendung des Summenscores.

Und Kubinger (2000) ergänzt als praktische Konsequenz: „dass Tests, die nicht gleich bei ihrer Entwicklung dem Rasch-Modell gemäß konstruiert wurden, letztlich unfair messen, läßt sich regelmäßig nachweisen: zum Beispiel für Ravens berühmten Matrizentest …". „Darauf hinzuweisen ist, dass die dem Rasch-Modell angepaßten Tests fast definitionsgemäß auf Leistungstests beschränkt (sind). Persönlichkeitsfragebogen sind wohl kaum modellkonform zu konstruieren: Die eigentlich interessierende Eigenschaft wird stets von einer zweiten, nämlich der, „wahr zu antworten", überlagert, so daß alle eindimensional messenden Modelle scheitern müssen."

Mangelnde Passung kann durch unterschiedliche Verstöße gegen die Voraussetzungen des Rasch-Modells verursacht werden:

- durch heterogene Stichproben. Für diesen Fall wurde u. a. die latente Klassenanalyse entwickelt. Sie unterteilt die Fähigkeitsdimension in Bereiche (Klassen), in denen ähnliches Antwortverhalten auftritt. Die Klassenanzahl schätzt die dafür zusätzlich erforderlichen freien Parameter und die Klassenanzahl. Alternativ könnte auch der Personenfit beurteilt werden.
- durch heterogene Items. Diese würden unterschiedliches Antwortverhalten hervorrufen und damit gegen die Annahme der (eindimensionalen) Skalierbarkeit verstoßen. Geprüft werden kann dies mit Homogenitätstests (z. B. der Mokkenanalyse oder Formen der Clusteranalyse).

– durch Verstoß, dass der Summenwert eine erschöpfende Statistik ist. Für diesen
Fall hat Birnbaum (vgl. Lord und Novick, 1968, S. 400) als Weiterentwicklung
ein zweiparametriges Modell entwickelt, in dem der gewichtete Summenscore
$t_j = \sum w_i \cdot a_{ji}$ als erschöpfende Statistik vorausgesetzt wird:

$$P \{a_{ij}=1 \mid f_j\} = (\exp[w_i (f_j - s_i)]) / (1 + \exp[w_i(f_j - s_i)]).$$

Die bisherigen Betrachtungen sind auch auf mehralternative Antwortmuster (z. B.
im multiple choice) erweiterbar. Dazu werden die Antwortwahrscheinlichkeiten
für jede Antwortalternative (also nicht nur für das Lösen) betrachtet. Schematisch
zeigt dies die folgende Abbildung 3 für ein Item mit drei Antwortkategorien [z. B.
0 (trifft nicht zu), 1 (trifft etwas zu), 2 (trifft zu)]:

Abbildung 3. Rasch-Modell Anwendung

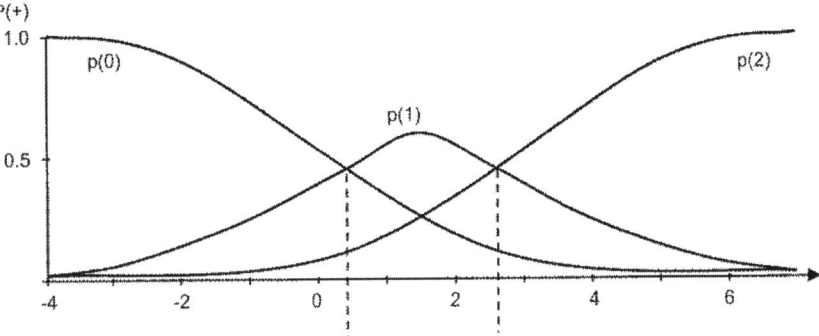

Entscheidend werden dann die Grenzen zwischen den einzelnen Antwortkategorien, häufig als Schwellen bezeichnet. Sie sind durch die Schnittpunkte der
Itemfunktionen ausgewiesen, da sich dort der Wechsel des Antwortverhaltens von
einer in die nächste Kategorie zeigt. Dieser Übergang wird durch sog. Schwellen-
funktionen q(x) beschrieben, die im mehrkategoriellen Fall folgende Form haben:

$$q(x) = (\exp(f - \tau_x)) / (1 + \exp(f - \tau_x)).$$

Dies ist wieder eine logistische Funktion, wobei die Schwellen τ_x dann jeweils
bei p= 0,5 liegen.

Für uns wichtig ist, dass man durch zusätzliche Annahmen die Freiheit der Modellparameter Parameter einschränken kann, so z. B. die Annahme der Äquidistanz
aller Schwellenabstände, womit die Existenz einer Metrik überprüft werden kann.
Die Parameterzahl reduziert sich bei dieser Annahme auf den einen Abstandspa-
rameter und je Item einen Lageparameter auf der latenten Dimension. Es gilt: je
restriktiver das Modell wird desto schwieriger ist eine ausreichende Passfähigkeit

zu sichern. Dennoch ist dies wohl der einzig gangbare Weg, um die Verwendung von Summenscores bei Ratingskalen wissenschaftlich zu rechtfertigen.

3.3 Zuverlässigkeit ist eine wesentliche Kenngröße psychologischer Methoden und wird im Sinne der KTT als Varianzverhältnis der aufgeklärten wahren Varianz zur Gesamtvarianz verstanden. Dies kennzeichnet gleichzeitig den Einfluss des Zufallsfehlers auf die Beobachtungen:

$$\text{Zuverlässigkeit} = (\text{Var}(T)) / (\text{Var}(X)) = 1 - \text{Var}(E)/\text{Var}(X) = \rho^2 (X,T) = \rho(X,X')$$

wobei der letzte Term die Möglichkeit der Schätzung der Reliabilität über einen parallelen Test X' kennzeichnet.

Häufig, insbesondere beim Vorliegen qualitativer Daten, wird zur Zuverlässigkeitskennzeichnung auch der Cronbach-Alpha-Koeffizient (Cronbach, 1951) verwendet, dessen Aussagekraft in den letzten Jahren problematisiert wurde. Historisch hat er mit der Kuder-Richardson-Formel K20 für dichotome Variable einen Vorgänger (1937) und auch Gutmann (1945) hatte schon fünf Schätzer vorgestellt, von denen lambda-3 mit alpha identisch ist. Cronbach (1951) weist selbst darauf und auch die Beziehung zur Formal von Spearman und Brown (für parallele Teile) hin und diskutiert deren Beziehungen.

Die theoretische Grundaussage zum Alpha-Koeffizienten formulieren Lord und Novick (1968) im Theorem 4.4.3. (S. 88): Wenn Y_i (i=1,…m) m Messungen mit den wahren Werten T_i sind und X deren Summe ist ($X = Y_1 + … + Y_m$), dann gilt:

$$\rho^2 (X,T) = (\sigma^2(T)) / (\sigma^2(X)) = \Sigma\sigma^2(T_i) / (\sigma^2(X)) \geq m/(m-1) \cdot [1 - (\Sigma\sigma^2(Y_i) / (\sigma^2(X))] = \alpha$$

(Der Beweis erfolgt mittels der Cauchy-Schwartz-Ungleichung.) Die Gleichheit gilt genau dann, wenn die Messungen Y_i t-äquivalent mit unabhängigen Fehlern sind, d.h. dass es für alle i, j eine Konstante a_{ij} so gibt, dass $T_i = a_{ij} + T_j$ ist. Dies bedeutet gleichzeitig, dass $\sigma^2 (T_i) = \sigma^2 (T_j)$ ist. Ansonsten entsteht die Frage nach der Einschätzung der Abweichung und den Konsequenzen für die Interpretation von alpha.

Sijtsma (2009) ist diesen Konsequenzen nachgegangen, indem er aus der Literatur bekannte alternative Beziehungen zur Schätzung der Reliabilität untersuchte: Neben alpha (Guttman's lambda-3) wurden betrachtet: a) die größte untere Schranke der Reliabilität glb (graetest lower bound) (vgl. Woodhouse und Jackson, 1977) und b) Guttman's lambda-2 als größere untere Schranke der Reliabilität (Guttman hatte insgesamt fünf untere Schranken eingeführt).

Für diese Schranken konnte theoretisch die folgende Beziehung ausgewiesen werden:

$$\text{alpha} \leq \text{lambda-2} \leq \text{glb} \leq \rho(X,X') = \rho^2(X,T).$$

Für die drei Schätzer der Reliabilität gilt die Gleichheitsbeziehung genau dann, wenn die Items essentiell t-äquivalent sind. Abweichungen von dieser Voraussetzung sollten sich also in den Unterschieden zwischen den Schätzern und den Konsequenzen für die Aussagekraft zeigen.

Für einen realen Datensatz wurden diese Schätzer bestimmt. Der Test (zum Copingverhalten) bestand aus 8 Ratingskalen (0, 1, 2, 3), die insgesamt eine zweifaktorielle Struktur aufwiesen. Diese ermöglichte es zusätzlich, den Test nach der Ladung in beiden Faktoren zu teilen. Auch für diese Testteile wurden die Koeffizienten bestimmt. Aus der vergleichenden Diskussion leitet Sijtsma u. a. folgende Konsequenzen ab: a) von allen unteren Schranken ist der Schätzer glb der beste und der einzige, der einen realistischen Wert ergibt. Sowohl alpha als auch lambda-2 können deutlich abweichen; and b) „Alpha is not a measure of internal consistency. Neither is it a measure of the degree of unidimensionality."

Letztere Aussage gilt genauso wie die Feststellung, dass die Reliabilität als Maß der Zuverlässigkeit weder ein Index der internen Konsistenz noch der Eindimensionalität ist. Reliabilität im Sinne der klassischen Testtheorie ist eine Abschätzung der Wirkung des Zufallsfehlers auf die Beobachtungsdaten, sie ist zwar Voraussetzung der Validität, hat aber selbst keinerlei Bezug zur inhaltlichen Struktur der Beobachtung. Dies gilt wohl auch für die vorab diskutierten Kennwerte, die allesamt formaler Natur und ohne Inhaltsbezug sind. So können hohe alpha-Koeffizienten sowohl bei ein- wie zweidimensionalen Messungen auftreten. Ich will abschließend darauf verweisen, dass es neuere Studien gibt, die die Mittel der Strukturgleichungsmodelle nutzen, um die Robustheit der Aussagen von Cronbachs-Alpha zu beurteilen (z. B. Maydau-Olivares u. a. 2010, Gu u. a. 2013).

4 Methodische Konsequenzen

a) Die methodischen und messtheoretischen Grundprobleme für die qualitative Persönlichkeitsforschung sind nicht weniger geworden und bestimmen weiterhin die Zuverlässigkeit ihrer Aussagen.

b) Für einige Grundprobleme haben die methodischen Entwicklungen jedoch neue Einsichten und auch neue Möglichkeiten erschlossen. Dies betrifft insbesondere

   – die Entwicklung differenzierterer Modellansätze (und damit auch Theorien), die Eigenschaften wie Eindimensionalität, Unabhängigkeit und auch Fehlereigenschaften zu modellieren und zu beurteilen gestatten,

   – die Entwicklung komplexerer Messstrukturen, die es gestatten, das Beziehungsgefüge von latenten und manifesten Variablen differenzierter zu

erfassen und damit den Verhältnissen im Phänomenbereich mehr zu entsprechen,
- die zunehmenden Versuche, die Robustheit von Modellen hinsichtlich von Verletzungen ihrer Voraussetzungen zu untersuchen und so deren Nutzung unter abgeschwächten Bedingungen zu rechtfertigen.

5 Kontakt

Prof. Dr. Bodo Krause
Institution: Humboldt-Universität zu Berlin, Institut für Psychologie
Address: Rudower Chaussee 14, 12489 Berlin, Germany
E-mail: bkrause@cms.hu-berlin.de

6 Literatur

Amthauer, R. (1973). *Intelligenzstrukturtest I-S-T 70*. Göttingen: Hogrefe.

Bargmann, R. (1954). *Signifikanz-Untersuchungen der Einfachen Struktur in der Factoren Analyse. Mitteilungsblatt für Mathematische Statistik*. Physica Verlag, Würzburg.

Bortz, J., & Döring, N. (2006). *Forschungsmethoden und Evaluation*. Heidelberg: Springer Medizin Verlag.

Cohen, J. (1988). *Statistical power analysis for behavioral sciences*. Hillsdale, New Jersey: Lawrence Erlbaum Ass.

Der Spiegel, Nr. 42/1978, S. 265–270.

Cronbach, L. J. (1951). *Coefficient alpha and the internal structure of tests. Psychometrika, 16*(3), 297–332.

Gigerenzer, G. (1981). Implizite Persönlichkeitstheorien oder quasi-implizite Persön-lichkeitstheorien? Eine Begriffsklärung und eine Validitätsstudie zu individuellen impliziten Theorien [Implicit personality theories or quasi-implicit personality theories? A concept clarification and a validity study on individual implicit theories]. *Zeitschrift für Sozialpsychologie, 12*, 65–80.

Gigerenzer, G. (2007) *Bauchentscheidungen. Die Intelligenz des Unbewussten und die Macht der Intuition*. Bertelsmann, München: C. Bertelsmann Verlag.

Gigerenzer, G. (2013). *Risiko. Wie man die richtigen Entscheidungen trifft*. München: C. Bertelsmann Verlag.

Gu, F., Little, T. D., & Kingston, N. M. (2013). Misestimation of reliability using coef-ficient alpha and structural equation modelling when assumptions of tau-equivalence and uncorrelated errors are violated. *Methodology, 9*(1), 30–40.

Guttman, L. J. (1945). A basis for analyzing test-retest reliability. *Psychometrika*, *10*(4), 255–282.

Horn, W. (1983). *L-P-S Leistungsprüfsystem*. Göttingen: Hogrefe.

Konitzer, F. (2013). Einmal ist keinmal. *Bild der Wissenschaft Band, 8*, 66–69.

Krause, B., & Metzler, P. (1978). Zur Anwendung der Inferenzstatistik in der psychologischen Forschung. *Zeitschrift für Psychologie, 186*, 244–267.

Krause, B., & P, Metzler (1983). *Angewandte Statistk*. Berlin: Deutscher Verlag der Wissenschaften.

Krause, B. (2007). *Zur Konstruktion von messtheoretisch begründeten Ratingskalen*. *Empirische Evaluationsmethoden* Bd. 11, 69–84. ZeE-Verlag: Berlin.

Krause, B. (2009). *Zur praktischen Bedeutsamkeit der Modellierung mit latenten Variablen*. *Empirische Evaluationsmethoden* Bd. 13, 27–38. ZeE-Verlag: Berlin.

Krause, B. (2014). *Zur Replikation wissenschaftlicher Befunde: Forschungsmethodische Aspekte zur aktuellen öffentlichen Diskussion*. In: *Empirische Evaluationsmethoden*, Bd. 18, 73–81. ZeE-Verlag: Berlin.

Kubinger, K. D. (2000). Replik auf Jürgen Rost „Was ist aus dem Rasch-Modell geworden?": Und für die Psychologische Diagnostik hat es doch revolutionäre Bedeu-tung. *Psychologische Rundschau, 51*, 33–40.

Kuder, G. F., & Richardson, M. W. (1937). The theory of estimationof test reliability. *Psychometrika, 2*, 151–160.

Lord, F. M., & Novick, M. R. (1968). *Statistical theories of mental test scores*. Reading, MA: Addison-Wesley.

Maydeu-Olivares, A., Coffman, D. L., Garcia-Ferero, C., & Gallardo-D. (2010). Hypothesis testing for coefficient alpha: An SEM approach. *Behavior Research Methods, 42*(2), 618–625.

Metzler, P., & Krause, B. (1997). Methodischer Standard bei Studien zur Therapieevaluation. In: Krause, B. & Metzler, P. (Hrsg.). Veränderungsmessung und Interventionsevaluation. *Empirische Evaluationsmethoden, Band 1*, 43–59. ZeE-Verlag: Berlin.

Metzler, P., & Krause, B. (1998). *Methodischer Standard bei Studien zur Therapieevaluation*. MPR Online, Bd. 2.

Rasch, G. (1960). *Probabilistic models for some intelligence and attainment tests*. (Copenhagen, Danish Institute for Educational Research), expanded edition (1980) with foreword and afterword by B. D. Wright. Chicago: The University of Chicago Press.

Rohrmann, B. (1978). Empirische Studien zur Entwicklung von Antwortskalen für die sozialwissenschaftliche Forschung. *Zeitschrift für Sozialpsychologie, 9*, 222–245.

Rost, J. (1999). Was ist aus dem Rasch-Modell geworden? *Psychologische Rundschau, 50*(3), 140–156.

Rost, J. (2003). *Zeitgeist und Moden empirischer Analysemethoden*. Forum Qualitative Sozialforschung/Forum Qualitative Social Research 4(2) Art 5 http://nbn-resolving.de/um:nbn.de.0114-fqs030258

Schramm, S. (2013). *Ein einmaliges Ergebnis.* http://www.zeit.de/2013/22/sozialpsychologische-studien

Spearman, C. (1904). General intelligence objectively determined and measured. *American Journal of Psychology, 15*, 201–293.

Tenzer, E. (2013). BIG FIVE unter Beschuss. *Bild der Wissenschaft, 7*, 72–76.

Thurstone, L. L. (1931). Multiple factor analysis. *Psychological Review, 38*, 406–427.

Thurstone, L. L. (1938). *Primary and mental abilities.* Chicago.

Überla, K. (1968). *Faktorenanalyse.* Berlin-Heidelberg-New York: Springer-Verlag.

Wright, S. S. (1921). Correlation and causation. *Journal of Agricultural Research, 20*, 557–85.

Hamidreza Khankeh[1,2] & Maryam Ranjbar[3,4]

[1] University of Social Welfare and Rehabilitation Sciences, Iran

[2] Karolinska Institute, Sweden

[3] Institute of Humanities and Social Studies, Iran

[4] University of Social Welfare and Rehabilitation, Iran

Conducting Qualitative Research in Health

Abstract. Qualitative research focuses on social world and provides investigators with the tools to study health phenomena from the perspective of those perceiving and experiencing them. This kind of research needs researchers bring their whole self into the process. To do this kind of research, identifying the research problem, forming the research question and selecting an appropriate methodology and research design are some of the initial challenges that researchers encounter in the early stages of a qualitative research project. These problems are common particularly for novice researchers. The purpose of this paper is to describe the practical challenges of qualitative inquiry in health and the challenges researchers faced doing interpretive research. The papers try to address the practical challenges of employing qualitative research in the field of health. This is a conceptual paper discussing the practical challenges to do qualitative research in the field of health, based on professional experience as a qualitative designer and available literature. One of the main topics discussed in this paper is to understand the nature of qualitative research, its inherent challenges and how to overcome them. Some of the challenges highlighted in this paper include: identifying the research problem, the formation of the research question/aim and selecting an appropriate methodology and research design, which are the main concerns of qualitative researchers and should be handled properly. Insights from the real-life experiences in conducting a qualitative research in the health reveal these practical issues. The paper provides personal commentaries on the experiences of a researcher in conducting purely qualitative study in field of health. It offers insights into practical difficulties encountered when performing qualitative studies and offers a glance into solutions and alternatives incorporated by the researcher, which could be of use to health care researchers.

Keywords: qualitative research, research methods, methodological challenges.

1 Introduction

Health services and health policy research can be developed through qualitative research methods of research, especially when it deals with a rapid change and develops a more fully integrated theory base and research agenda. However, the field must be put up with the best traditions and techniques of qualitative methods and should distinguish the essentiality of special training and experience in applying these methods (Sofaer, 1999).

Qualitative research methodologies could help us improve our understanding of health related phenomena. Health knowledge must also include interpretive action to maintain scientific quality when research methods are applied. Qualitative and quantitative strategies should be seen as complementary rather than thought as incompatible. Although the procedures of interpreting texts are different from those of statistical analysis, due to their different type of data and questions to be answered, the underlying principles are much the same (Malterud, 2001). During more than a decade working as qualitative designer, I have faced with a lot of challenges in conducting qualitative research in the field of health which occupied the mind of health researchers. This article then contributes to the discussion of challenges related to qualitative research in healthcare in the light of personal experiences of a researcher in conducting an academic purely qualitative study regarding health.

2 Literature

2.1 A main issue for the qualitative researcher

Qualitative research methods involve systematic collection, organizing, and interpretation of material in textual form derived from talk or observations. They are useful to explore the meanings of social phenomena as experienced by individuals themselves, in their natural context. The health community still looks at qualitative research with skepticism and accuses it for the subjective nature and absence of facts. Scientific standards and criterias do exist nevertheless the adequacy of guidelines has been vigorously debated within this cross-disciplinary field (Malterud, 2001).

Clinical knowledge consists of interpretive action and interaction–factors that involve communication, shared opinions, and experiences. The current quantitative research methods indicate a confined access to clinical knowledge, since they insert only the questions and phenomena that can be controlled, measured, and are countable. Where it is necessary to investigate, share and contest the tacit knowing of an experienced practitioner. Qualitative research focuses on the people's social world, and not their disease. It is concerned with increasing understanding the meaning of certain conditions for health professionals and patients, and how their relationships are built in a particular social context (Ceña, Rodríguez, & Martinez, 2012). These kinds of research allow the social events explored as experienced by individuals in their natural context. Qualitative inquiry could contribute to a broader understanding of health science (Malterud, 2001) considering the substantial congruence between the core elements of health practice and the principles underpinning qualitative research.

In this paper we are going to discuss important practical challenges of qualitative inquiry in health and the challenges faced by researchers to use interpretive research.

2.2 Why qualitative research in the health professions?

Researcher should justify the reason he or she selected qualitative research. Qualitative researchers pursue a holistic and exclusive perspective. The approach is helpful in development of understanding human experiences, important for health professionals who focus on caring, communication and interaction (Holloway & Wheeler, 2013). Many potential researchers intend to find the answer to the questions about a problem or major issue in clinical practice or education for practice that cannot be verified by quantitative research.

In fact they choose qualitative research for some significant reasons:

- The emotions, perceptions and actions of people who suffer from a medical condition can be understood by qualitative research.
- The meanings of health professions will only be uncovered through observing the interactions of professionals with clients and interview about their experience. This is also applicable to the students destined for the healthcare field.
- Qualitative research is individualized; hence researchers consider the participants as whole human beings not as a bunch of physical compartments.
- Observation and asking people are the only ways to understand the causes of particular behaviors. Therefore, this type of research can develop health or education policies; policies for altering health behavior can only be effective if the behavior's basis is clearly understood (Holloway & Wheeler, 2013).

2.3 Choosing an approach for health research

Researchers select approaches and methodology based on some scientific logics not being easy or interesting. The nature and type of the research question or problem, the researcher's epistemological stance, capabilities, knowledge, skills and training and the resources available for the research project are the criteria which adopting methodology and procedures depend on (Corbin & Strauss, 2008; Holloway & Wheeler, 2013). Inconsistency between research question and methodology, insufficient methodological knowledge and lack of attention on philosophical underpinning of qualitative methodology can be mentioned as some important challenges here. There are several different ways of qualitative research and researchers will have to select between varieties of approaches. The qualitative research is based on the theoretical and philosophical assumptions that researchers try to understand. Then research methodology and process should be chosen consistent with these basic assumptions and the research question as well (Holloway & Wheeler, 2013).

2.4 Research question and aim

Qualitative research is exciting because it asks questions about people's everyday lives and experiences. You will have the chance of discovering the "significant truths" in lives of people as a qualitative researcher. That is a wonderful privilege but you need to get those questions right if you dig into people's lives and ask about their real experiences. An adequate and explicit research question, or set of interrelated questions, builds the basis for a good research. But excellent research questions are not easy to write at all. A good research requires a good research question as well because it allows us to identify what we really want to know. However, at the beginning of a project, researchers may be uncertain about what exactly they intend to know, so vague questions can lead to an unfocused project.

Common problems coming up with a research question include:

– Deciding about the research area among a range of issues that heeded in your field of interest.
– Not capable of pointing towards any interesting area or topic sufficient to focus a major piece of work on.
– Knowing about the area you want to concentrate on but not a certain topic.
– Knowing what area and topic which is specifically difficult to articulate a clear question.

Just make sure that you give serious consideration to the chosen area as the basis of your research and that a qualitative project is relevant and possible.

Having identified a research area, your next step will be to identify a topic within that interesting area. Research questions should be derived from the literature. Based on my experience, novice researchers have some problems find the right topics in their field of interest because they do not do a broad literature review to find the gaps and problems suitable to be investigated. Sometimes their field of interest is different from their supervisors or there are no experts to help them in this regard.

Although the topic may retain your interest and you may be committed to undertake such a study, it is important to recognize that some topics of personal relevance may also be deeply significant and difficult to research. Finally you need to make sure that your topic of interest is the one that you can actually study within the project constraints such as time and fund (Kinmond, 2012).

Once you have identified your interesting topic for research (according to a broad literature review, personal and professional experience and/or expert opinion) then you can begin to create research question.

Forming the research question is one of the initial challenges that researchers encounter in the early stages of a research project. Therefore it acquires the significance by the very fact that it provides brief, but nevertheless, important information on the research topic that allows the reader to decide if the topic is relevant, researchable and a remarkable issue. Furthermore, the research question in qualitative studies has an additional significance as it determines the manner of conducting the study.

The content of a good qualitative research question takes the form of a declarative rather than an interrogative statement. For instance: Exploring the experiences of self-immolated women regarding their motives for attempting suicide: A qualitative content analysis study in Kermanshah Iran. Make sure that your research question is consistent with the approach you are adopting. It is like an easy trap if you decide about the research question before considering the proper way you are intending to make assumptions and analyze your data.

My experiences showed that novice researchers formulate their research question without considering the approach of their study in a proper way and usually their research questions are very broad, unclear and vague. Since the intention of their studies are not completely clear at the beginning, they cannot decide about the research approach and they have to change their research question and take different directions in the course of study or they will end up without adequate results that can help readers or consumers improve their understanding or solve the problem.

2.5 Choosing right methodology and research design

Crucial decisions need to be made about an appropriate methodology, such as ethnography or grounded theory after identifying the initial research question. The main concern of novice researchers is to find the reason and appropriate design to do the research, and proper methodology to answer the question. Researchers ought to figure out about the planning of qualitative research and how to choose the methodology.

The research design and methodology must be adequate to the selected topics and to the research question. Researchers have to identify, describe and justify the methodology they chose besides the strategies and procedures involved. Then it is pivotal to find the proper method for the research question. It should be noticed that some of the details of a qualitative research project cannot be ascertained in advance and may be specified as they arise during the research process (Holloway & Wheeler, 2013).

Little acknowledgement about different approaches address different kinds and levels of questions and take a different stance on the kind of phenomena that it is focused upon is an important problem for novice researchers. More discussion and debate are necessary before selecting and justifying an approach.

The need for consistency and coherence becomes more obvious when we consider the risk of something called 'method-slurring'. This is the problem of blurring distinctions between qualitative approaches. Each approach has to demonstrate its consistency to its foundations and will reflect them in data collection, analysis and knowledge claim.

3 Discussion

Qualitative research focuses on social world and provides investigators with the tools to study health phenomena from the perspective of those experiencing them. Identifying the research problem, forming the research question and selecting an appropriate methodology and research design are some of the initial challenges that researchers encounter in the early stages of a qualitative research project. Once the research problem and initial research question identified, the crucial decisions will be to select the appropriate methodology. Subsequent arrangements would be on the proper methods of data collection, choosing participants and the research setting according to the methodology and to the research question. It is highly recommended that researchers have to exactly understand the nature and character of their inquiries and of the knowledge they choose to create before adhering to a distinct research methodology based on scientific knowledge. As the final word, the researcher should make sure that he/she gives serious consideration to the chosen area as the basis of research and that a qualitative project is relevant and possible. Then forming research question in proper way and selecting appropriate methodology can guarantee original, interesting and applied knowledge which at least can increase our understanding about the meaning of certain conditions for professionals and patients; and how their relationships are built in a particular social context.

4 Affiliations

Dr. Hamidreza Khankeh
Institution 1: University of Social Welfare and Rehabilitation Sciences, Tehran, Iran
Address: kodakyar Ave., daneshjo Blvd., Evin, Post code:1985713834
Institution 2: Karolinska Institute
Address: Södersjukhuset (KI SÖS), Sjukhusbacken 10, 118 83 Stockholm, Sweden
E-mail: hamid.khankeh@ki.se

M.A. Maryam Ranjbar
Institution: University of Social Welfare and Rehabilitation, Institute of Humanities and Social Sciences, and Social Determinant of Health Research Center
Address: Enghelab St. Tehran, Iran
E-mail: maryam.ranjbar.75@gmail.com

5 References

Ceña, D. P., Rodríguez, J. M., & Martínez, E. P. (2012). Emergency nursing (2): qualitative research in emergency medicine; design and areas of applications in emergency care. *Emergencias, 24*, 410–413.

Corbin, J., & Strauss, A. (2008). *Basics of qualitative research: Techniques and procedures for developing grounded theory*. California, London, New Delhi, Singapore: Sage.

Holloway, I. (2005). *Qualitative research in health care*. Maidenhead, England: McGraw-Hill International.

Holloway, I., & Wheeler, S. (2013). *Qualitative research in nursing and healthcare*. Chichester, West-Sussex: John Wiley & Sons.

Kinmond, K. (2012). Coming up with a research question. In C. Sullivan, S. Gibson, & S. Riley (Eds.), *Doing your Qualitative Psychology Project* (pp. 23–36). London: Sage.

Malterud, K. (2001). The art and science of clinical knowledge: Evidence beyond measures and numbers. *The Lancet, 358*(9279), 397–400.

Malterud, K. (2001). Qualitative research: standards, challenges, and guidelines. *The Lancet, 358*(9280), 483–488.

Sofaer, S. (1999). Qualitative methods: What are they and why use them? *Health services research, 34*, 1101.

Yumi Lee[1] & Thérèse Thuemler[2]

[1] University of Leipzig, Germany

[2] Martin Luther University Halle-Wittenberg, Germany

A Content-analysis of Korean and German Teachers' Perception and Belief regarding Students with ADHD: A Comparison with US Findings

Abstract. The purpose of this study was to understand Korean and German teachers' perceptions and beliefs regarding children with ADHD and to make a comparison with two previous US studies (Blume-D'Ausilio, 2005; Whitworth, Fossler, & Harbin, 1997). Two open-ended statements were adapted from a study conducted by Whitworth et al. (1997). A content-analysis was used to analyze a total of 240 Korean and 104 German primary school teachers who had previously taught children with ADHD. Teachers' responses were thoroughly reviewed and coded. Triangulation method was used for validity and the differently classified responses were re-evaluated and appropriately modified. Both Korean and German teachers' perceptions were similar, which are noticeably different to those from two US studies. Regarding teachers' beliefs, three countries showed similar findings. Comparison of these findings suggests that ADHD Management Manual should be developed based on specific cultural context. It could help teachers not only to understand the student with ADHD but also to establish the confidence to manage the student with ADHD. Additional research for pre-service teachers also needs to be explored in order to establish successful management of the student with ADHD when they become a teacher.

Keywords: ADHD, culture, primary school, open-ended statement, triangulation method.

1 Introduction

Since the majority of children with ADHD attend regular schools in Korea and in Germany, it is no exaggeration to say that the success of children with ADHD is mainly based on how teachers manage their problematic behaviors as well as their personal, emotional, and social needs in the classroom. Therefore, it is imperative for teachers to be well prepared by having a comprehensive understanding about ADHD, developing favorable attitudes towards children with ADHD, and applying their understanding to specific management strategies in the classroom. In reality, however, teachers often face difficulties to effectively manage children with

ADHD in their classroom. They feel frustrated, pessimistic, and overwhelmed about managing children with ADHD due to their problematic behavioral characteristics (e.g., not listening to teachers instructions, frequently fighting with peers) in comparison to typically developing children (Greene, Beszterczey, Katzenstein, Park, & Goring, 2002; Hong, 2008).

Numerous studies found that teachers hold significantly less favorable attitudes (e.g., Jeong & Choi, 2010), have a lack of knowledge about ADHD (e.g., Schmiedeler, 2013), and less confident to manage these children (e.g., Lee, 2015). These teachers' stress may influence their perceptions and expectations towards children with ADHD in a negative way (e.g., negative perceptions and expectations) (e.g., Bussing, Gary, Leon, Garvan, & Reid, 2002), and may in turn affect children's performance, behavior, and self-esteem (e.g., Bekle, 2004). Therefore, it is necessary for teachers to be well prepared to provide children with ADHD an appropriate care in school settings (e.g., Emmer, Evertson, & Worsham, 2003; Holz & Lessing, 2002).

The purpose of this study was to understand Korean and German teachers' perceptions and beliefs regarding children with ADHD and to make a comparison with two previous US studies. Following research questions (RQ) were addressed by the study. What perceptions and beliefs do Korean and German primary school teachers have regarding children with ADHD? Will teachers' perceptions and beliefs be significantly different between Korea and Germany?

2 Method

2.1 Sample

240 Korean primary school teachers (grade 1–6) and 104 German primary school teachers (grade 1–4) teachers participated in this study.

2.2 Survey instrument

Two open-ended statements were adapted from a study conducted by Whitworth et al. (1997) as follows:

(a) perceptions: "The most difficult thing about teaching children with ADHD is…",
(b) beliefs: "I believe that I would be more successful teaching children with ADHD if….".

2.3 Data analysis

A content-analysis was used. Regarding the first statement (perceptions), a total of 144 responses from 94 answered Korean teachers and 82 responses from 60 German answered teachers were analyzed. As for the second statement (beliefs), a total of 94 responses from 75 answered Korean teachers and 74 responses from 58 answered German teachers were analyzed. Responses were thoroughly reviewed and coded based on modified criteria from previous two US studies by first and second authors (see Table 1). Triangulation method was used based on coding criteria by all authors in order to proof validity. The differently classified responses were re-evaluated and appropriately modified.

Table 1. Coding criteria

Perceptions	Beliefs
• Teacher time/attention/Energy • Teacher/ self-improvement • Behavior management • Keeping them focused/paying attention • Staying on task /work completion • Lack of organization skills • Lack of social skills • Lack of parent/home support • Accommodating needs • No problem • Miscellaneous	• More training/workshop • Smaller class size • Parent support • Support/resources • Aid/volunteer • Early diagnosis • Medication • Miscellaneous • Teachers' characteristic

Note. Whitworth et al. (1997)'s criteria was slightly modified by first and second authors.

3 Results

3.1 "The most difficult thing about teaching children with ADHD is..."

As shown in Table 2, teachers' perceptions were similar in Korea and Germany[1], which are noticeably different to those from two US studies[2] (Blume-D'Ausilio, 2005; Whitworth et al., 1997). Both Korean and German teachers' perceived that (a) *behavior management* (e.g., disruptions to class and other children) (Korea: 38.8 %; Germany: 35.5 %), (b) *teachers self-improvement* (e.g., to control my anger

---

1 Contact the first author for Korean and German teachers' responses.
2 See Blume-D'Ausilio (2005, pp. 168–174) for US teachers' responses to the first open-ended statement.

and staying calm, no knowledge) (Korea: 16.7 %; Germany: 29.3 %), and (c) *accommodating needs* (e.g., a big class, other children' understanding) (Korea: 9.7 %; Germany: 6.1 %) were the most difficult thing about teaching children with ADHD.

Table 2. "The most difficult things about teaching children with ADHD…"

	Lee & Thuemler (2015)		Blume-D'Ausilio (2005)
	Korea	Germany	US
1	Behavior management (38.8 %)	Behavior management (35.5 %)	Keeping them focused (28.5 %)
2	Teacher/self-improvement (16.7 %)	Teacher/self-improvement (29.3 %)	Behavior management (23.1 %)
3	Accommodating needs (9.7 %)	Accommodating needs (6.1 %)	Off task/completion (13.8 %)

On the other hand, previous US study (Blume-D'Ausilio, 2005) found that the most difficulties about teaching children with ADHD were (a) *keeping them focused* (e.g., they have a difficult time focusing/paying attention) (28.5 %), (b) *behavior management* (e.g., controlling them in a classroom setting. Keeping them in check so as not to disturb the rest of the class) (23.1 %), and (c) *staying on task/work completion* (e.g., keeping them focused and on task with the rest of the children) (13.8 %).

3.2 "I believe that I would be more successful teaching children with ADHD if…"

As shown in Table 3, teachers' beliefs were similar in Korea and Germany[3], which were also similar to the founding of two US studies[4] (Blume-D'Ausilio, 2005; Whitworth et al., 1997). Both Korean and German teachers believed that they would be more successfully teaching children with ADHD if they had (a) *more training/workshop* (e.g., good quality training) (Korea: 42.6 %; Germany: 27.5 %), (b) *smaller class size* (e.g., fewer children per class) (Korea: 14.9 %; Germany: 27.0 %), and (c) *support/resources* (e.g., less administrative work) (Korea: 13.8 %; Germany: 26.0 %).

---

3 Contact the first author for Korean and German teachers' responses.
4 See Blume-D'Ausilio (2005, pp. 175–180) for US teachers' responses to the second open-ended statement.

Table 3. "I believe that I would be more successful teaching children with ADHD if..."

	Lee & Thuemler (2015)		Blume-D'Ausilio (2005)
	Korea	Germany	US
1	More training/workshop (42.6 %)	More training/workshop (27.5 %)	More training/workshop (55.5 %)
2	Smaller class size (14.9 %)	Smaller class size (27.0 %)	Smaller class size (11.9 %)
3	Support/resources (13.8 %)	Support/resources (26.0 %)	Aid/Volunteer (9.6 %)

Similarly, Blume-D'Ausilio (2005) found that US teachers believed that they would be more successful to teach children with ADHD if (a) they had *more training* (e.g., I was able to attend workshops that had ADHD children or parents and past teachers of ADHD students) (55.5 %), (b) there were *smaller classes* (e.g., the number of children in the class would be kept at 15–18) (11.9 %), and (c) they had *an aid/ a volunteer* (e.g., there was help in the classroom to manage difficult behaviors when instruction is taking place) (9.6 %), which were also similar to Whitworth et al. (1997).

4 Discussion

By comparing the results of different countries it could be shown that teachers, independently from cultural background, are struggling with ADHD-children. Especially, they ask for further information about this disorder and looking for strategies how to manage these children during the lessons. Therefore, it is necessary to develop a specific manual-based training, where all needed information are taught and culturally necessary skills are practiced, in order to better prepare the teachers and to reduce their stress. This could help teachers not only to change their previous perceptions and beliefs about children with ADHD establish their confidence to manage them, but also, and maybe the most important thing, to help those children to be more successful in school. Additional research for pre-service teachers also needs to be explored in order to establish successful management of the student with ADHD when they become a teacher.

5 Affiliations

Dr. Yumi Lee
Institution: University of Leipzig, Educational and Rehabilitation Psychology
Address: Neumarkt 9–19, 04109 Leipzig, Germany
E-mail: yumi.lee@uni-leipzig.de

Dipl.-Psych. Thérèse Thuemler
Institution: Martin Luther University Halle-Wittenberg, Educational Science
Address: Franckeplatz 1, 06110 Halle (Saale)
E-mail: therese-nicola.thuemler@paedagogik.uni-halle.de

6 References

Bekle, B. (2004). Knowledge and attitudes about attention-deficit hyperactivity disorder (ADHD): a comparison between practicing teachers and undergraduate education students. *Journal of attention disorders, 7*(3), 151–161.

Blume-D'Ausilio, C. (2005). *Sources of information and selected variables and their relationship to teachers' knowledge and attitudes regarding attention deficit hyperactivity disorder (ADHD)* (Unpublished doctoral dissertation). Florida Atlantic University Boca Raton, Florida, USA.

Bussing, R., Gary, F. A., Leon, C. E., Garvan, C. W., & Reid, R. (2002). General classroom teachers' information and perceptions of attention deficit hyperactivity disorder. *Behavioral Disorders, 27*(4), 327–339.

Emmer, E. T., Evertson, C. M., & Worsham, M. E. (2003). *Classroom management for secondary teachersrs*. Boston: Allyn and Bacon.

Greene, R. W., Beszterczey, S. K., Katzenstein, T., Park, K., & Goring, J. (2002). Are students with ADHD more stressful to teach?: Patterns of teacher Stress in an elementary school sample. *Journal of Emotional and Behavioral Disorders, 10*(2), 79–89.

Holz, T., & Lessing, A. (2002). Reflections on attention-deficit hyperactivity disorder (ADHD) in an inclusive education system: Research paper. *Perspectives in Education, 20*(3), 103–110.

Hong, Y. (2008). Teachers' perceptions of young children with ADHD in Korea. *Early Child Development and Care, 178*(4), 399–414.

Jeong, J. S., & Choi, J. O. (2010). An examination of elementary teachers' knowledge of ADHD, attitudes toward including children with ADHD, and use of behavior management strategies. *The Journal of Special Education: Theory and Practice, 11*(3), 371–393.

Lee, Y. (2015). *Teachers' attitudes, knowledge, and classroom management strategies for students with ADHD: a cross-cultural comparison of teachers in South-Korea and Germany* (Unpublished doctoral dissertation). University of Leipzig, Leipzig, Germany.

Schmiedeler, S. (2013). Wissen und Fehlannahmen von deutschen Lehrkraeften ueber die Aufmerksamkeitsdefizit-/Hyperaktivitaetsstoerung (ADHS), *Psychologie in Erziehung und Unterricht, 60*(2), 143–153.

Whitworth, J. E., Fossler, T., & Harbin, G. (1997). Teachers' perceptions regarding educational services to students with attention deficit disorder. *Rural Educator, 19*(2), 1–5.

Beiträge zur Pädagogischen und Rehabilitationspsychologie
Studies in Educational and Rehabilitation Psychology

Herausgegeben von / Edited by Evelin Witruk

www.peterlang.com